Soldier Protective Clothing and Equipment

Feasibility of Chemical Testing Using a Fully Articulated Robotic Mannequin

Committee on Full-System Testing and Evaluation of Personal Protection Equipment Ensembles in Simulated Chemical-Warfare Environments

Board on Chemical Sciences and Technology

Division on Earth and Life Studies

NATIONAL RESEARCH COUNCIL
OF THE NATIONAL ACADEMIES

THE NATIONAL ACADEMIES PRESS
Washington, D.C.
www.nap.edu

THE NATIONAL ACADEMIES PRESS 500 Fifth Street, N.W. Washington, DC 20001

NOTICE: The project that is the subject of this report was approved by the Governing Board of the National Research Council, whose members are drawn from the councils of the National Academy of Sciences, the National Academy of Engineering, and the Institute of Medicine. The members of the committee responsible for the report were chosen for their special competences and with regard for appropriate balance.

This study was supported by the U.S. Department of Defense under Grant W911NF-06-C-0176. Any opinions, findings, conclusions, or recommendations expressed in this publication are those of the authors and do not necessarily reflect the views of the organizations or agencies that provided support for the project.

International Standard Book Number-13: 978-0-309-10933-8
International Standard Book Number-10: 0-309-10933-7

Additional copies of this report are available from the National Academies Press, 500 Fifth Street, NW, Lockbox 285, Washington, DC 20055; (800) 624-6242 or (202) 334-3313 (in the Washington metropolitan area); Internet, http://www.nap.edu.

Cover: Design by Michael Dudzik; photos courtesy of U.S. Army.

Printed in the United States of America

THE NATIONAL ACADEMIES

Advisers to the Nation on Science, Engineering, and Medicine

The **National Academy of Sciences** is a private, nonprofit, self-perpetuating society of distinguished scholars engaged in scientific and engineering research, dedicated to the furtherance of science and technology and to their use for the general welfare. Upon the authority of the charter granted to it by the Congress in 1863, the Academy has a mandate that requires it to advise the federal government on scientific and technical matters. Dr. Ralph J. Cicerone is president of the National Academy of Sciences.

The **National Academy of Engineering** was established in 1964, under the charter of the National Academy of Sciences, as a parallel organization of outstanding engineers. It is autonomous in its administration and in the selection of its members, sharing with the National Academy of Sciences the responsibility for advising the federal government. The National Academy of Engineering also sponsors engineering programs aimed at meeting national needs, encourages education and research, and recognizes the superior achievements of engineers. Dr. Charles M. Vest is president of the National Academy of Engineering.

The **Institute of Medicine** was established in 1970 by the National Academy of Sciences to secure the services of eminent members of appropriate professions in the examination of policy matters pertaining to the health of the public. The Institute acts under the responsibility given to the National Academy of Sciences by its congressional charter to be an adviser to the federal government and, upon its own initiative, to identify issues of medical care, research, and education. Dr. Harvey V. Fineberg is president of the Institute of Medicine.

The **National Research Council** was organized by the National Academy of Sciences in 1916 to associate the broad community of science and technology with the Academy's purposes of furthering knowledge and advising the federal government. Functioning in accordance with general policies determined by the Academy, the Council has become the principal operating agency of both the National Academy of Sciences and the National Academy of Engineering in providing services to the government, the public, and the scientific and engineering communities. The Council is administered jointly by both Academies and the Institute of Medicine. Dr. Ralph J. Cicerone and Dr. Charles M. Vest are chair and vice chair, respectively, of the National Research Council.

www.national-academies.org

COMMITTEE ON FULL-SYSTEM TESTING AND EVALUATION OF PERSONAL PROTECTION EQUIPMENT ENSEMBLES

MASAYOSHI TOMIZUKA (*Chair*), University of California, Berkeley
HADI ABU-AKEEL, AMTENG Corporation, Sterling, Virginia
CHRISTOPHER G. ATKESON, Carnegie Mellon University, Pittsburgh, Pennsylvania
LISA M. BROSSEAU, University of Minnesota, Minneapolis
ZANE FRUND, Mine Safety Appliances Company, Pittsburgh, Pennsylvania
DARRELL L. JAN, Jet Propulsion Laboratory, California Institute of Technology, Pasadena
SUNDARESAN JAYARAMAN, Georgia Institute of Technology, Atlanta
LEO KOBAYASHI, Rhode Island Hospital Medical Simulations, Providence
DAVA J. NEWMAN, Massachusetts Institute of Technology, Cambridge
ARTHUR C. SANDERSON, Rensselaer Polytechnic Institute, Troy, New York
R. PAUL SCHAUDIES, GenArraytion Inc., Rockville, Maryland

Staff

DOROTHY ZOLANDZ, Director, Board on Chemical Sciences and Technology
KATHRYN HUGHES, Postdoctoral Fellow
TINA M. MASCIANGIOLI, Responsible Staff Officer
KELA MASTERS, Project Assistant
EMILY ANN MEYER, Program Officer, National Materials Advisory Board
JESSICA PULLEN, Research Assistant
NORMAN GROSSBLATT, Senior Editor

BOARD ON CHEMICAL SCIENCES AND TECHNOLOGY

Acknowledgment of Reviewers

This report has been reviewed in draft form by persons chosen for their diverse perspectives and technical expertise in accordance with procedures approved by the National Research Council's Report Review Committee. The purpose of this independent review is to provide candid and critical comments that will assist the institution in making the published report as sound as possible and to ensure that it meets institutional standards of objectivity, evidence, and responsiveness to the study charge. The review comments and draft manuscript remain confidential to protect the integrity of the deliberative process. We wish to thank the following for their review of this report:

Mr. Geoff Betsinger, 3M, St. Paul, Minnesota
Dr. C.W. Cole, Clemson University, Pendleton, South Carolina
Dr. William R. Hamel, University of Tennessee, Knoxville
Dr. Duane Linder, Sandia National Laboratories, Livermore, California
Dr. Dennis J. Paustenbach, ChemRisk, Inc., San Francisco, California
Dr. Jimmy L. Perkins, University of Texas Health Science Center, San Antonio
Dr. Matthew Reed, University of Michigan, Ann Arbor
Dr. Mike Sailor, University of California, San Diego
Dr. Esther S. Takeuchi, State University of New York, Buffalo
Mr. Joseph J. Vervier, ENSCO, Inc., Melbourne, Florida

Although the reviewers listed above provided many constructive comments and suggestions, they were not asked to endorse the conclusions or recommendations nor did they see the final draft of the report before its release. The review of this report was overseen by Dr. Frederick A. Murphy, University of Texas Medical Branch, Galveston, and Dr. Robert A. Beaudet, University of Southern California, Emeritus. Appointed by the National Research Council, they were responsible for making certain that an independent examination of this report was carried out in accordance with institutional procedures and that all review comments were carefully considered. Responsibility for the final content of this report rests entirely with the authors and the institution.

Preface

About a year ago—the U.S. Department of Defense, Joint Program Executive Office for Chemical and Biological Defense, Joint Project Manager, Nuclear, Biological and Chemical Contamination Avoidance, Product Director, Test Equipment, Strategy, and Support (PD TESS)—met and initiated this study with National Research Council staff. The study evaluates the feasibility of developing a fully articulated robotic mannequin to test individual protection ensembles (IPE) in chemical-warfare agent environments, namely the Protection Ensemble Test Mannequin (PETMAN) system. PD TESS was seeking assistance from an objective group of scientists and engineers (see Appendix A for the statement of task) who could evaluate the technical merit and feasibility of the PETMAN system requirements (listed in Appendix B).

The resulting study was performed by an ad hoc committee with expertise in chemistry; chemical engineering; biology; human physiology; chemical sensing; respiratory protective equipment; materials science; robotics; articulated mannequins; cost, benefit, and risk analysis; and warfare simulation involving chemical agents. Committee members were sought from various sources, including the academic, national laboratory, and industrial sectors.

A committee of 10 members met in person four times from January to May 2007 to discuss overall U.S. Department of Defense (DOD) and specific PD TESS objectives in protection ensemble testing. During that time, the committee received briefings on current capabilities (see Appendix D), deliberated, and developed its conclusions and recommendations. In the report, the committee discusses in detail what it considered to be the key design challenges associated with developing a PETMAN system: human-physiology simulation, chemical-agent sensing, robotics design, "skin" architecture and materials, and systems integration.

I thank all the members of the committee for their contributions to this study. Briefings by guest speakers were extremely useful in helping the committee to understand some critical aspects of the PETMAN requirements. On behalf of the committee, I thank the staff members of the Board on Chemical Sciences and Technology who helped us by organizing the meetings and making our service a rewarding and enjoyable experience.

Masayoshi Tomizuka, *Chair*
Committee on Full-System Testing and Evaluation of Personal Protection Equipment Ensembles in Simulated Chemical-Warfare Environments

Contents

Summary

As part of the continued need to protect soldiers in environments where they are exposed to chemical-warfare agents, the effectiveness of personal protective equipment must be ensured by testing. However, carrying out tests with human subjects presents numerous challenges. A wide array of humanoid robots and mannequins are increasingly used in national-security applications. Over the years, human-size thermal mannequins have been used to test military garments, and there is growing interest in making mannequins more human-like by adding motion to permit more advanced testing capabilities. The U.S. Department of Defense (DOD) and its counterparts in other countries have been pursuing the development of such systems for many years. For example, the United Kingdom has developed Portonman, and similar systems exist in the Netherlands and Canada. However, those systems and other U.S. efforts fall short of meeting the testing needs described by the U.S. Department of Defense, Joint Program Executive Office for Chemical and Biological Defense, Joint Project Manager, Nuclear, Biological and Chemical Contamination Avoidance, Product Director, Test Equipment, Strategy, and Support (PD TESS) requirements (see Appendix B).

This report addresses the feasibility of developing an advanced humanoid robot—the Protection Ensemble Test Mannequin (PETMAN) system—to enhance the testing of chemical-warfare personal protective equipment in the United States. The PETMAN system performance requirements include both *threshold* and *objective* requirements, where a threshold requirement is a "must have" while an objective requirement denotes a "would like to have" operational capability. Although most of the PETMAN system requirements taken independently are technically feasible, fulfilling all of PD TESS requirements in a single PETMAN (especially at the objective level) is not currently possible. The option of using a tethered mannequin significantly increases the feasibility of creating a PETMAN system. This report discusses variations on meeting the threshold and objective requirements and a complementary approach to the PETMAN system that might be adequate to meet DOD's short term testing needs.

STUDY BACKGROUND

PD TESS has identified the need for full-system testing and evaluation of individual protective ensembles (IPE) in chemical-warfare agent environments and seeks to develop a PETMAN system to meet the need. PD TESS envisions the PETMAN system as a fully articulated robotic mannequin that will perform exercises that simulate war-fighter activities. The performance requirements set by PD TESS also call for the PETMAN system to be heated to produce body temperatures and to be able to perspire and breathe. It must be constructed of materials that will not be substantially degraded by exposure to chemical agents and that can be decontaminated to negligible levels of chemical contaminants without adverse affects on its operation.

PD TESS requested that the National Academies assemble a committee to assess the feasibility of PETMAN system according to detailed performance requirements. The study task was specified as follows:

- Determine the feasibility of a PETMAN system, based on all delineated PD TESS system design requirements for such a system.
- Focus on the significant design challenges associated with the PD TESS PETMAN system and whether and how they might be addressed.
- Discuss the cost-benefit and risk-benefit trade-offs associated with various design approaches to a PETMAN.
- Discuss whether and how some or all of the necessary protective ensemble test capability could be obtained if a PETMAN capability is infeasible. Discuss the cost-benefit and risk-benefit trade-offs of these alternatives.

The study was performed by an ad hoc committee consisting of persons with expertise in chemistry, chemical engineering, biology, human physiology, chemical sensing, respiratory protective equipment, materials science, robotics, articulated mannequins, cost-benefit and risk-benefit analyses, and warfare simulations involving chemical agents. These experts were sought from various sources and include people in the academic, national laboratory, and industrial sectors.

A committee of 10 met in person four times in January-May 2007, and received briefings on current capabilities, deliberated, and determined its conclusions and recommendations. PD TESS asked that the study conclusions and recommendations be submitted by August 2007.

PETMAN SYSTEM REQUIREMENTS

As discussed earlier, PD TESS envisions the PETMAN system as a fully articulated robotic mannequin that will perform exercises that simulate war-fighter activities. PD TESS has developed detailed performance requirements for PETMAN, as provided in Appendix B, and summarized in the text that follows. Again, the PETMAN system performance requirements include both threshold (T) and objective (O) requirements, where a threshold requirement is a "must have" while an objective requirement denotes a "would like to have" operational capability. If a requirement is not identified by (T) or (O), then it is considered to be a threshold requirement. While PD TESS desires a PETMAN system meeting all the objective-level requirements, a PETMAN system meeting the threshold-level requirements is also a suitable option for PD TESS.

Is compatible with individual protective equipment. The system should be compatible with all individual protection and ancillary equipment weapon systems, including the ability to hold, grip, and aim a weapon. The system design must also meet the appropriate 50th percentile human male measurements. Ideally, the system would be compatible with the use of Skin Exposure Reduction Paste against Chemical Warfare Agents (SERPACWA) (O).

Is not tethered. Ideally, the PETMAN system will be free-standing and self-contained (O)—that is, there will be no external support, and all power, fluids, heating, and other components for operation will be contained internally. A tethered system design (T) is an option, but the tether must not compromise the integrity of the IPE equipment.

Uses off-the-shelf technology and can be decontaminated. The PETMAN system should use as many common, commercially available components as possible. The materials of construction cannot be substantially degraded by exposure to traditional chemical agents (T) and ideally will be resistant to toxic industrial chemicals (TICs) and toxic industrial materials (TIMs) (O). It should be able to be decontaminated to negligible levels without adverse effects on the operation of the PETMAN system (T).

Operates continuously for 12-24 hours. The system ideally will be capable of operating for 24 hours before requiring operational maintenance, 6 months before preventive maintenance, and 12 months before calibration (O). At a minimum, the PETMAN system should be capable of operating for 12 hours before requiring operational maintenance, 3 months before preventive maintenance, and 6 months before calibration (T).

Tests for agent in real time. Ideally, the system will be designed to enable integration with real-time (1-s increments) sampling technologies, procedures, and equipment and to record the following system characteristics over time: skin temperature, respiration rate, perspiration rate, and total mass (in nanograms) of chemical vapor that penetrates the IPE (O). At a minimum, the system must be compatible with current chemical-agent sampling technologies, procedures, and equipment (T).

Simulates human physiology. The system will simulate realistic variable (O) or fixed (T) human skin temperature, perspiration rate (variable at 0.11-1.8 L/h or fixed at 0.4 L/h), and respiration rate (tidal volume of 10-115 L/min and variable breath frequency or fixed tidal volume of 1.5 L and breath frequency of 33 breaths/min).

Is compatible with test-chamber conditions. The system must be able to operate at: a temperature of 90°F ± 2°F (T); –25°F to 125°F ± 1°F, measured every 5 min (O); a relative humidity of 80% ± 3% (T); 0-100% ± 1%, measured every 5 min (O); a wind speed of 0-10 mph ± 10% (T); 0-161 mph ± 2 mph (O); and 0.25 iwg chamber vacuum maintained ± 2%. In addition, the system will need to be compatible with the use of liquid and vapor chemical agents including: all nerve and vesicant agents, as well as the chemical simulants, triethylphosphate and methyl salicylate.

Is compatible with Man-in-Simulant Test exercises. The system must be articulated and robotic so that it can simulate the Man-in-Simulant Test (MIST) exercises, which include standing, walking, marching, doing modified jumping jacks, kneeling, crawling, and holding a weapon (T) (Figure S.1). The system must also be programmable to perform a series of exercises or motions and be able to track body position during all motions in 1-s increments (T). Ideally, the system would be able to perform a full jumping jack, which involves simultaneously bringing hands and feet together and apart (O).

Has human-like articulation and construction. Finally, the system should be articulated and robotic so that it looks and moves like a human, including

FIGURE S.1 Man-in-Simulant Test exercises in the chamber. The PETMAN system is required to perform the same set of exercises as the soldiers shown.
SOURCE: Charles Walker, Dugway Proving Ground, U.S. Department of Defense.

aesthetics, proportions, and how the joints respond to sudden movements (T). All movements must simulate realistic human control.

A substantial effort will be needed to move from current systems, such as Portonman, to the type of system described in the PETMAN requirements document. This report addresses the feasibility of meeting the major design challenges of the PETMAN system requirements on the basis of currently available technology. This includes a detailed consideration of human-physiology simulation, the mannequin under ensemble sensing conditions, robotic design, architecture and materials of the PETMAN skin, an integrated PETMAN system, and a complementary approach to PETMAN. The overarching conclusions and recommendations of the report are provided below.

OVERARCHING CONCLUSIONS AND RECOMMENDATIONS

PD TESS should consider the following three overarching issues as it seeks to develop a PETMAN system: priority setting among the PETMAN requirements, contractor qualifications, and a complementary option to a PETMAN system. Each is discussed in the following sections.

Priority Setting Among PETMAN Requirements

The PETMAN requirements cover all the desired and required features of mannequin function with little or no priority setting. PD TESS indicated that a concurrent exercise was being carried out to set priorities among the requirements but that the results would probably not be available in time to inform the PETMAN feasibility study. This feasibility study concludes the following:

Conclusion 1: Taken independently, most of the PETMAN threshold requirements can be met with current technologies or incremental developments of existing technologies (see detailed discussions in Chapters 2 through 6).

Conclusion 2: Several options are available for chemical-agent sensing, robotic actuation, and overall system design.

Conclusion 3: Using currently available technologies, it may be possible to meet some of the threshold requirements in a nontethered system, but largely due to available battery technology such a system would be limited to an operating time of two hours. The other threshold and objective requirements may be difficult or impossible to meet with a nontethered mannequin.

Conclusion 4: Integrating all the current objective requirements will be a major challenge for design and implementation of a full PETMAN system.

Conclusion 5: Meeting the threshold requirement of a tethered system, which would reduce the number of subsystems housed in the mannequin, is feasible. However, design challenges still exist for incorporating all necessary systems into a single PETMAN.

In light of those conclusions, PD TESS should take the following actions:

Recommendation 1: To achieve greater success with the future proposal solicitation for a PETMAN system, PD TESS should set priorities among the PETMAN requirements according to the program objectives.

Recommendation 2: PD TESS should adopt a phased approach to the development of the PETMAN system, first addressing the high-risk design challenges identified in the study (see detailed discussion in Chapter 6) and then deciding on the achievable objectives according to the priorities it has set.

Contractor Qualifications

The development of a PETMAN system is a large undertaking for any organization. The development of individual components—in particular the robotics capability—will require considerable resources and expertise. Simulating human physiology in addition to developing a robot may be beyond the means of a single group. On the basis of that assessment, the study concludes the following:

Conclusion 6: The design and development of a PETMAN system will require a multidisciplinary effort that encompasses expertise in computer software engineering, robot design, mannequin design, materials science and engineering, human physiologic simulation, sensor technologies, and systems integration.

PD TESS should take the following action in connection with assessing the qualifications of the company or companies chosen to develop a PETMAN system:

Recommendation 3: The primary contractor should have demonstrated capabilities in systems integration.

Recommendation 4: A workshop should be organized to inform the proposing groups fully of the objective and threshold requirements. The invitation list should include system integrators and developers and suppliers of component technologies for the mannequin, materials, and sensors.

A Complementary Approach

The current Man-in-Simulant Test (MIST) protocol evaluates individual protection ensembles (IPE) on soldiers rather than mannequins, and this offers the benefit of testing the effects of actual human movements and physiologic conditions. However, one of the major shortcomings of MIST is its method of under-ensemble data acquisition. It has been shown (see chapters 2, 5, and 7) that recreating human-like movement, respiration, perspiration, and body proportions will be difficult and expensive. In contrast, simulant chemicals mimic almost all the physical features of actual chemical agents. On the basis of that assessment, the study concludes the following:

> **Conclusion 7:** Some of the technologies reviewed in this report—such as real-time sensing of chemicals, temperature, and humidity—could be used in the MIST to provide real-time leak detection and characterization of the microenvironments in the protective garments.

> **Conclusion 8:** Some technologies are sufficiently mature to support construction of a whole-body suit for a human—a sensor-integrated body suit (SIBS)—outfitted for real-time sensing of chemicals, body temperature, heart rate, cardiographic characteristics, and humidity without the need for a tether (see Chapter 7 for discussion of the SIBS). Such a suit would allow substantial improvement in the MIST without the expense and risk associated with a fully developed PETMAN system and at a small fraction of costs in money and time. The only limitation would be the inability to use actual chemical agents.

In light of those conclusions, PD TESS should take the following action:

Recommendation 5: A SIBS should be seriously considered as an update of the MIST and as complementary to the proposed PETMAN system. Unless there is an absolute requirement of all the capabilities associated with the proposed PETMAN system, this sensor-integrated approach would provide many of the key capabilities in the interim. Such an approach would provide substantial improvements over current testing while the critical paths and absolute requirements for PETMAN development are explored. It would allow testing to continue with human subjects and allow collection of data on a broader array of human characteristics than will be possible with the PETMAN system.

In light of the full list of requested functionalities of the PETMAN system, it will be difficult to design a nontethered, free-standing, 50th-percentile-male robotic test mannequin that can operate continuously beyond the available battery capacity of two hours. Tests longer than two hours will require reduced motion, a recharging or refueling method, a tether, or other approaches that would require research extending beyond the sponsor's desired development timeframe. The main design challenge for the PETMAN system is the integration of all system components (power, control, sensors, perspiration, respiration, actuation, and so on) within the size constraints. Setting priorities among the PETMAN systems requirements, improving guidance for proposing contractors, and considering complementary test approaches with simulants and real-time sensing will enhance the ability of PD TESS to develop a PETMAN system and ultimately improve the protection capability of IPE against chemical-warfare agents.

1

Introduction

The protection of soldiers in environments where they are exposed to chemical-warfare agents remains a need of the U.S. Department of Defense (DOD), and it is necessary to test protective equipment for effectiveness under such conditions. Ideally, DOD would like to have the capability for testing and evaluating (T&E) individual protection ensembles (IPE) with real chemical agents in authentic war-fighter environments—which would fill a gap in IPE testing and evaluation (Figure 1.1). Since it is not possible to use real chemical agents on human test subjects, such DOD efforts as the Man-in-Simulant Test (MIST) perform full-system IPE T&E with chemical simulants. DOD also performs T&E on IPE components with chemical agents, but this does not fulfill its desire for full-system testing.

To avoid the use of human subjects, testers are increasingly moving toward the use of a wide array of robots and mannequins. For example, human-size thermal mannequins have been used to test soldier uniforms and protective garments. There is also growing interest in making mannequins more human-like by adding realistic motion and other human characteristics, such as body temperature and sweating, to provide more advanced testing capabilities. Examples of those efforts are discussed briefly in this chapter and in more detail later (see chapters 2 and 4). To determine the feasibility of developing an advanced humanoid robot—the Protection Ensemble Test Mannequin (PETMAN)—to enhance chemical-warfare IPE testing, PD TESS requested that the National Research Council conduct a study that would:

- Determine the feasibility of a PETMAN system on the basis of all delineated PD TESS design requirements for such a system.
- Focus on the significant design challenges associated with the PD TESS PETMAN system and whether and how they might be addressed.
- Discuss the cost-benefit and risk-benefit trade-offs associated with various design approaches to a PETMAN system.
- Discuss whether and how some of or all the necessary protection ensemble test capability could be obtained if a full PETMAN system were infeasible and discuss the cost-benefit and risk-benefit trade-offs of alternatives.

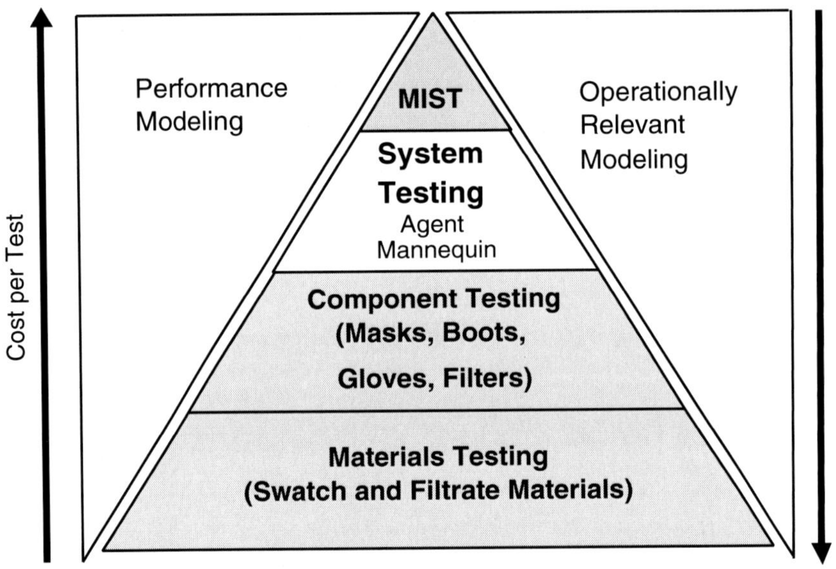

FIGURE 1.1 Chemical Agent Testing and Evaluation Pyramid for IPE.
SOURCE: Charles Walker, Dugway Proving Ground, U.S. Department of Defense.

THE PROTECTION ENSEMBLE TEST MANNEQUIN SYSTEM

The purpose of developing a PETMAN system is to test the protection capability of IPE against chemical-warfare agents. The PETMAN system is envisioned as a fully articulated robotic mannequin that will perform exercises that simulate war-fighter activities, that will be heated to produce human body temperatures, and that will be able to perspire and breathe. A descriptive overview of the desired features of the PETMAN system design is provided in the text that follows. The significant design challenges are provided in Table 1.1, and a detailed list of system requirements is provided in Appendix B. The PETMAN system performance requirements, however, include both *threshold* and *objective* requirements, where

TABLE 1.1 Summary of PETMAN System Significant Design Challenges 3.2.1-3.2.8 (see Appendix B)

Design Challenge	Threshold Requirement	Objective Requirement
3.2.1	• Tethered	• Freestanding
3.2.2	• Compatible with all individual and ancillary equipment and weapons systems. • Meet 50th percentile male anthropometric measurements	• Compatible with skin exposure reduction paste against chemical-warfare agents
3.2.3	• Not degraded by exposure to traditional chemical agents • Capable of being decontaminated with no adverse effects	• Not significantly degraded by exposure to TICS and TIMS • Capable of being decontaminated such that there is negligible agent residue
3.2.4	• Compatible with current under-ensemble chemical breakthrough sampling technologies, procedures, and equipment • At a minimum, sampling locations shall be the same as those defined in TOP 10-22-022	• Designed to enable integration with real-time (1 second increments) sampling technologies, procedures, and equipment • Breakthrough measured in nanograms
3.2.5	• Fixed skin temperature, and perspiration and respirations rates by body region	• Realistic variability in skin temperature, and perspiration and respiration rates.
3.2.6	• Fixed environmental chamber conditions	• Range of environmental chamber conditions
3.2.7	• Can perform the Man-in-SimulantTest exercises versus all motions (partial jumping jacks)	• Full jumping jacks
3.2.8	• Minimum amount of hand and foot articulation	• Fully articulated hands and feet that simulate human motion

a threshold requirement is a "must have" while an objective requirement denotes a "would like to have" operational capability. While PD TESS desires a PETMAN system meeting all the objective-level requirements, a PETMAN system meeting the threshold-level requirements is also an option for PD TESS.

Compatibility with Existing Protective Equipment and Weapons Systems: Because the main objective of the PETMAN system is to test IPE, it must be compatible with individual protection and ancillary equipment (listed in Box 1.1), including donning and doffing and proper sizing and fit of the IPE. In particular, the PETMAN system must meet the appropriate 50th percentile male anthropometric measurements, as defined in DOD-HDBK-743A, *Military Handbook Anthropometry of U.S. Military Personnel* (13 February 1991), to allow for the necessary fit and seal that each piece of protective equipment requires. The PETMAN system must also be able to hold and aim designated weapon systems (requirement 3.3.10, Appendix B) in accordance with field manual (FM) 3-22.9, the *Rifle Marksmanship Field Manual.*

Mannequin Support: Ideally, the PETMAN system will be free-standing and self-contained. That is, it will have no external support, and all power, fluids, heating, and other components for operation will be contained internally. This could be accomplished with a tethered design with external supports, power requirements, and telemetry connections, but the use of a tether must not compromise the integrity of the IPE equipment being tested with the PETMAN system. If a tethered design is selected, it must not compromise the whole ensemble operation.

Materials of Construction and Decontamination: The PETMAN system should use as many commercially available components as possible. At a minimum, the PETMAN system materials of construction cannot be significantly degradable by exposure to traditional chemical agents. Ideally, the system will also be resistant to toxic industrial chemicals (TICs) and toxic industrial materials (TIMs). It should be possible to decontaminate the system to negligible levels of chemicals without adversely affecting its operation, as defined by the 3X decontamination level in DA PAM 385-61, *Toxic Chemical Agent Safety Standards.*

Operational Time: The PETMAN system ideally will be capable of operating for 24 hours before operational maintenance, for 6 months before preventive maintenance, and for 12 months before calibration. At a minimum, it should be capable of operating for 12 hours before operational maintenance, for 3 months before preventive maintenance, and for 6 months

BOX 1.1
Individual Protection and Ancillary Equipment, with Which a PETMAN System Should Be Compatible (See Requirement 3.3.9, Appendix B)

Suits
Joint Service Lightweight Suit Technology (JSLIST) Type II
JSLIST Type VII
All Purpose-Personal Protective Ensemble (AP-PPE)
Chemical Protective Undergarment (CPU)

Boots
Green Vinyl Overboot (GVO) / Black Vinyl Overboot (BVO)
Multipurpose Overboot (MULO)

Gloves
7, 14, and 25 mil butyl gloves
JSLIST Block 1 Gove Upgrade (JB1GU) and JB1GU-Flame Retardant (FR)

Masks
M40/42 Series Masks
M45 Mask
M48 Chemical-Biological Apache Aviator Mask
Aircrew Eye Respirator Protection (AERP)
AP-22P Respirator Assembly

Helmets
Advanced Combat Helmet (ACH)
Personal Armor System Ground Troops (PASGT) Helmet
Modular Integrated Communications Helmet (MICH)
Lightweight Helmet

Combat Boots
Lightweight Desert Combat Boot
Jungle Boot
Infantry Combat Boot
Temperate and Hot Weather Combat Boot

Ballistic Protection Vests
Spall Vest
Interceptor Vest

Pistol Holsters

All Services Battle Dress and Combat Uniform

All Services Physical Training Gear (T-shirt, running shorts, socks)

Skin Exposure Reduction Paste against Chemical Warfare Agents (SERPACWA)

before calibration. *Operational maintenance* is defined as the maintenance procedures required to prepare the PETMAN system for each test trial, for example, filling a perspiration reservoir, changing agent samplers, or decontamination before the next trial. *Preventive maintenance* is defined as maintenance performed before a failure to prevent its occurrence.

Sampling: The PETMAN system should be designed to enable integration with real-time (1-second increments) sampling technologies, procedures, and equipment and to record the following system measures over the desired operational time given above: skin temperature, respiration rate, perspiration rate, and total mass (in nanograms) of chemical vapor that penetrates or permeates the protection ensemble. At a minimum, the system must be compatible with current under-IPE sampling technologies, procedures, and equipment as defined in *Test Operations Procedure (TOP) 10-2-022, Chemical Vapor and Aerosol System-Level Testing of Chemical/ Biological Protective Suits.*

Human Simulation: The PETMAN system must simulate the following environmental and physiologic conditions under the IPE:

- **Temperature:** Realistic variability in body surface temperature based on body region and level of physical activity and exertion, or at a minimum a fixed skin temperature based on body region.
- **Perspiration:** Realistic variability in perspiration rates (range, 0.11-1.8 L/h) based on level of physical activity and exertion, or at a minimum a fixed perspiration rate of 0.4 L/h.
- **Respiration:** Realistic variability in respiration rate (range, 10-115 L/min with variable tidal volumes and breath frequencies) based on level of physical activity and exertion, or at a minimum a respiration rate of 50 L/min (fixed tidal volume of 1.5 L and breath frequency of 33/min).

Test-Chamber Conditions: The PETMAN system must be able to operate in the test chamber under various conditions given below. In addition, PD TESS has indicated that the PETMAN system will be required to operate in an agent chamber, which will be 8'×8'×10' (L×W×H). The dimensions of the current MIST chamber, are 10'×12'×8.5' (L×W×H).

- **Temperature:** The ideal temperature test range is –25°F to 125°F ± 1°F, measured every 5 min. At a minimum, the system must operate at 90°F ± 2°F.
- **Relative humidity:** The ideal relative-humidity test range is 0-100% ± 1%, measured every 5 min. At a minimum, the system must operate at 80% ± 3% relative humidity.
- **Wind speed:** The ideal wind-speed test range is 0-161 mph ± 2 mph. At a minimum, the system must operate at 0-10 mph ± 10%.
- **Pressure:** The pressure range is 0.25 ± 2% inches water gauge (iwg) chamber vacuum maintained.

- **The system must be able to operate in the presence of liquid and vapor chemical agents, including** all nerve and vesicant agents, and the chemical simulants triethylphosphate and methyl salicylate.

Compatibility with Man-in-Simulant Test Exercises: The PETMAN system must be articulated and robotic so that it can reproduce the Man-in-Simulant Test (MIST) exercises listed below. Figure 1.2 shows a schematic of a MIST chamber, and Figure 1.3 shows MIST exercises being performed in a chamber. The system must also be programmable to perform a series of exercises or motions and be able to track body position during these exercises and motions in 1-s increments.

- **Standing.**
- **Walking** at 4.8 km/h (3 mph).
- **Marching** (12-in. high step) at 4.8 km/h (3 mph).
- **Jumping jacks:**
  - Ideally, start with feet together and hands at sides. Simultaneously bring hands together palm to palm over head while jumping and landing with feet shoulder-width apart. Jump, simultaneously bringing feet back together and hands back to sides.

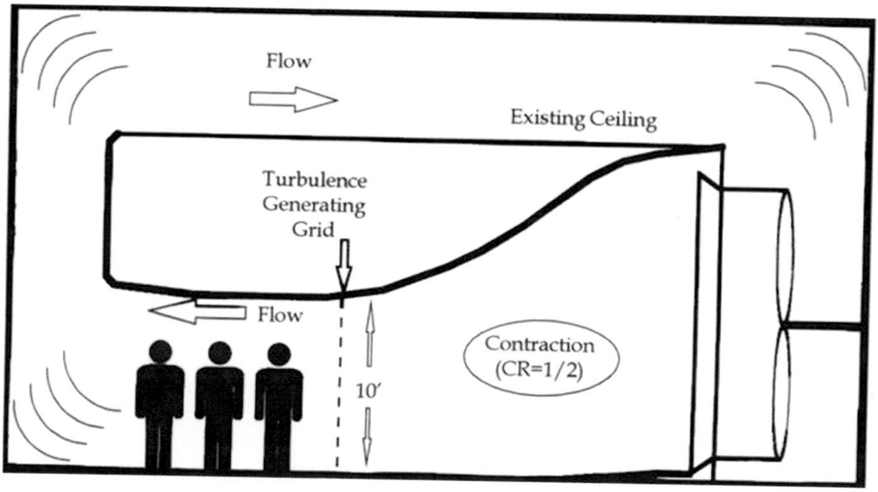

FIGURE 1.2 Schematic of a Man-in-Simulant Test chamber.
SOURCE: Charles Walker, Dugway Proving Ground, U.S. Department of Defense.

FIGURE 1.3 Man-in-Simulant Test exercises being performed in a chamber.
SOURCE: Charles Walker, Dugway Proving Ground, U.S. Department of Defense.

  - At a minimum, start with feet together and hands at sides. Simultaneously move left foot to the side, causing feet to be shoulder-width apart, and bring hands together palm to palm over head. Simultaneously return left foot to starting position (feet together) and hands back to sides. Repeat the exercise with the right foot.
- **Sitting.**
- **Moving from standing position to squatting position and returning to standing position.**
- **Reaching arms in all directions.**
- **Moving from standing to lying prone and returning to standing.**
- **Kneeling on one knee.**
- **Kneeling on both knees.**
- **Low crawling:** Lie prone. Keep body flat against the ground. With

firing hand, grasp weapon sling at the upper sling swivel. Let the front handguard rest on forearm (keeping the muzzle off the ground), and let the weapon butt drag on the ground. To move, push arms forward and pull firing-side leg forward. Then pull with arms and push with leg. Continue this throughout the move.

- **High crawling:** Keep body off the ground by resting on forearms and lower legs. Cradle weapon in arms and keep its muzzle off the ground. Keep knees well behind buttocks so body will stay low. To move, alternately advance right elbow and left knee, then left elbow and right knee.
- **Aiming weapon in various positions:** Standing, kneeling on one knee, lying prone (grip rifle, sight rifle, trigger pull associated with small arms).

Articulation and Construction: Finally, the PETMAN system should be articulated and robotic so that it looks and moves like a human, including aesthetics, proportions, and how the joints respond to sudden movements. All movements shall simulate realistic human control. The following are the minimal degrees of freedom (DoFs) and considerations for the PETMAN system to mimic human control. Additional DoFs or joints may be used to allow the PETMAN system to mimic human control and don and doff IPE as defined earlier under "Compatibility with Existing Protective Equipment and Weapons Systems."

- **Upper body:** Consists of a head, a neck, a torso, two shoulders, two upper arms, two elbows, two lower arms, and two hands.
- **Head:** Able to both pan and tilt as required.
- **Neck:** Provides at least two DoFs between the torso and the head. This will allow the head to look up and down and from side to side or to pitch and yaw.
- **Torso:** The base component of the upper body. The basic shape of the torso is similar to that of a male human chest. The torso has at least three DoFs: pitch, roll, and yaw. The PETMAN system is able to bend over, tilt from side to side, and swivel with respect to the frontal plane.
- **Shoulders:** Is attached to the torso and allows for at least two DoFs.
- **Upper arm:** Is attached at the shoulder and is designed so that it is able to move with at least two DoFs without restriction.
- **Elbow:** Has at least one DoF and provides an attachment between the upper and lower arms.
- **Lower arm:** Extends from the elbow joint and has a hand mechanism at its distal end.

- **Wrist:** Is able to move with at least two DoFs for wrist extension, flexion, and abduction.
- **Hand:** Is articulated to study the glove-coat interface or to have an opposable thumb and four fingers. Each finger has at least four DoFs. The distal interphalangeal joint and the proximal interphalangeal joint each have one DoF. The metacarpophalangeal joint has at least two DoFs due to flexion and abduction. The opposable thumb has at least three DoFs. The thumb has one DoF for the interphalangeal joint. The thumb metacarpophalangeal joint has at least two DoFs due to flexion and abduction. Additional thumb DoFs (an objective goal) will be due to flexion and abduction of the trapeziometacarpal joint.
- **Lower body:** Consists of the waist, two hip joints, two upper legs, two knees, two lower legs, two ankles, and two feet.
- **Hip joint:** Allows the upper leg to move with at least two DoFs.
- **Knee:** Moves with at least one DoF.
- **Ankle:** Moves with at least two DoFs and allows the foot to both pitch and roll.
- **Ball of foot:** Moves with at least one DoF, providing toe-roll motion while walking.
- **Joints:** Move smoothly and efficiently to mimic human motion. The PETMAN system movement is free of backlash.

EXAMPLES OF CURRENT MILITARY PETMAN-LIKE SYSTEMS

DOD and its counterparts in other countries have been pursuing the development of a PETMAN-like system for many years. Brief descriptions of the U.K. and Canadian systems based on publicly available information are provided below. Both fall short of meeting PD TESS requirements for PETMAN outlined earlier. First, the systems are not freestanding; they are attached to support systems, which limits the mannequin range of motion and affects compatibility with existing protective equipment and weapons systems. Second, there is no real-time chemical sampling or simulation of human physiology. The articulation and construction of these systems are inadequate for making it look or move like a human.

United Kingdom System

The Defense Science and Technology Laboratory, in Porton Down, United Kingdom has developed an articulated mannequin capable of typical ranges of human movement while wearing protective clothing. The Porton-man mannequin is based on 50th percentile male dimensions as determined from anthropometric surveys of 2,500 U.K. army and 2,500 U.K. naval

FIGURE 1.4 Portonman mannequin with knapsack sprayer in operation, simulating application of agricultural pesticides.
SOURCE: Research Report RR004: Dermal Exposure Resulting from Liquid Contamination 2002. ISBN 0717625303. Health and Safety Executive, United Kingdom. © Crown copyright material is reproduced with the permission of the Controller of HMSO and Queen's Printer for Scotland.

personnel. The mannequin is articulated at the shoulder, elbow, and hip and knee joints; movement of the limbs is in the vertical plane. Movement is achieved through linkages to the hands and feet, which are attached to a motor-driven pulley system that produces a normal or exaggerated marching motion. The relative radii of the pulleys govern the extent of limb movement. Cycle time is adjustable, although testing is normally conducted at 30 cycles/min, resulting in an equivalent walking speed of 5.4 km/h (3.4 mph), although this value was obtained when only the arms were set in motion.[1] Figure 1.4 shows Portonman being used to test for pesticide exposure. Figure 1.5 shows the under-ensemble sampling locations (test points).

[1]Defence Science and Technology Laboratory, 2002 Dermal exposure resulting from liquid contamination. Research Report 004 prepared for the Health and Safety Executive, United Kingdom, *www.hse.gov.uk*.

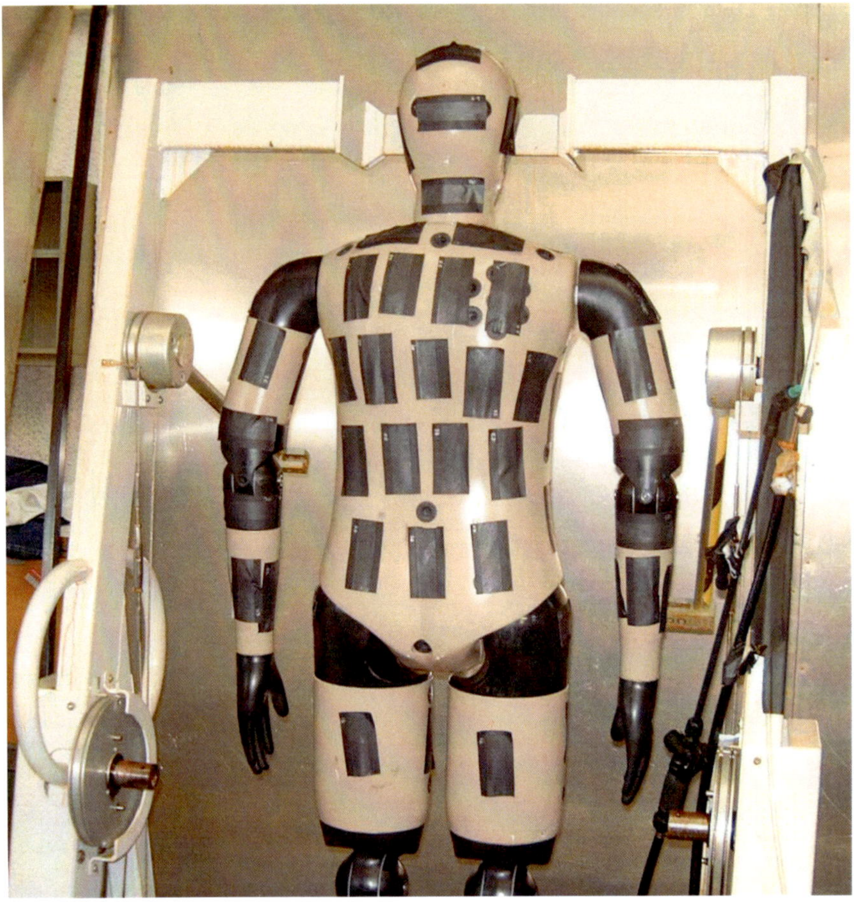

FIGURE 1.5 Portonman mannequin with sampling tapes mounted before dressing.
SOURCE: Research Report RR004: Dermal Exposure Resulting from Liquid Contamination 2002. ISBN 0717625303. Health and Safety Executive, United Kingdom. © Crown copyright material is reproduced with the permission of the Controller of HMSO and Queen's Printer for Scotland.

Canadian System

Defense Research and Development Canada Suffield has developed an articulated mannequin system designed and built by Crawley Creatures, U.K. The mannequin weighs 24-27 kg, has a shell made of carbon-fiber composite material, and is anthropometrically based on a 50th percentile Canadian military male.

Like Portonman, the Canadian mannequin is mounted on a support structure to accomplish human-like movements. It can walk, run, squat and lift, sit (with legs raised to a sitting position), and "creep" (walking in a crouching position). A separate articulated human head form is capable of simulating facial and neck movements of the human head.

The mannequin and support structure sit on a revolving platform that is integral to the test chamber. The support structure is electrically and physically connected to the test chamber. The system is computer-controlled and programmable, and a data-acquisition system records time-stamped mannequin-movement data.

ORGANIZATION OF THIS REPORT

In chapters 2 through 6, the key design challenges associated with developing a PETMAN system are considered in detail: human-physiology simulation, chemical-agent sensing, robot design, skin materials, and system integration. Chapters 2 through 5 include a discussion of current capabilities, feasibility, and design challenges associated with the PETMAN-system requirements and how they might be addressed, risk-benefit and cost-benefit analyses of requirements and trad-eoffs, and potential alternatives. A complementary approach to developing a PETMAN system is presented in Chapter 7, and overarching conclusions and recommendations are provided in Chapter 8.

Scope of Report

The presentation of technologies in this report is not comprehensive. The study task was to determine feasibility of the PETMAN requirements, not actually design the PETMAN system. In light of this task, some potential technologies and the thought process needed for considering and incorporating such technologies into a PETMAN system are presented. Anyone considering creating a PETMAN system will have to carry out a more thorough survey to determine the most appropriate technologies to use.

In addition, the rationale behind all the PETMAN requirements is not critically evaluated in this report, because the DOD sponsors indicated that a concurrent exercise was being carried out to set priorities among the requirements. However, the results of that exercise were not available in time to inform the PETMAN feasibility study. Questions did arise during the study about the need to test chemical agents rather than simulants. DOD explained that it needs to recreate an authentic war-fighter environment—which would be important for demonstrating test capability to soldiers.

2

Design Challenge: Simulation of Human Physiology

This chapter addresses the simulation of temperature, perspiration, and respiration, described in PETMAN design challenge 3.2.5 (see Appendix B):

The study will determine the feasibility of designing a PETMAN system that can simulate fixed skin temperature (by body region), perspiration rate (by body region), and respiration rate (threshold level) and a realistic variability in skin temperature, perspiration rates, and respiration rates based on the amount of physical activity/exertion (objective level) defined in 3.3.4.1-3.3.4.3.

This chapter includes a discussion of the relevant PETMAN requirements (Box 2.1), current technologic capabilities, design challenges and how they might be addressed with current and future technology, costs and benefits of different approaches, and feasibility and potential trade-offs.

CURRENT TECHNOLOGIC CAPABILITIES

The human-physiology simulation aspects of the PETMAN system may be best categorized as follows: temperature and perspiration, respiration and ventilation, and physiology integration. Each of those functions has been simulated to some degree by currently available systems; a few systems incorporate all three. We reviewed all currently available systems that can simulate at least one, and we include here information on technical specifications, images, and references.

BOX 2.1
Relevant PETMAN Requirements

The requirements below outline the desired level of simulating human physiology in the PETMAN system excerpted from the detailed list of PETMAN system design requirements that appear in Appendix B:

3.3.3 The PETMAN system shall meet the anthropometric requirements of the 50th percentile male in accordance with DOD-HDBK-743A, *Military Handbook Anthropometry of U.S. Military Personnel*, 13 February 1991.

3.3.4 The PETMAN system shall simulate the following environmental/physiological conditions under the individual protection ensemble.

3.3.4.1 The PETMAN system shall simulate fixed skin temperature by body region (T) and more realistic variability in body surface temperature based on body region and the level of physical activity/exertion (O).

3.3.4.2 The PETMAN system shall simulate a fixed perspiration rate of 0.4 L/hr (T) and more realistic variability in perspiration rates (range 0.11 to 1.8 L/hr) based on the level of physical activity/exertion (O).

3.3.4.3 The PETMAN system shall simulate a respiration rate of 50 L/min (fixed tidal volume of 1.5 L & breath frequency of 33 breaths/min) (T) and more realistic variability in respiration rates (range 10 to 115 L/min with variable tidal volumes and breath frequencies) based on the level of physical activity/exertion (O).

3.3.13 The PETMAN system shall record the following system parameters over time: skin temperature, respiration rate, perspiration rate, and total mass (in nanograms) of chemical vapor that penetrates/permeates through the protective ensemble. The PETMAN system shall record the start and stop time of each motion in 1 second increments.

NOTE: T=threshold (minimum requirement) and O=objective

Temperature and Perspiration

Several generations of thermal mannequins have been developed in the last 70 years for various commercial and military purposes. Used primarily for garment evaluation, early versions manufactured in the 1940s through 1960s were standing, mostly nonarticulated and nonperspiring models. Second-generation variants with movable body parts and appendages started to appear in the 1970s. The addition of perspiration capability has been more recent in what may be considered third-generation and later thermal mannequins. Several documents reviewing those types of mannequins (also referred to as manikins) and their evolutionary changes are available.[1]

Military Systems

Information on and specifications of several mannequins constructed by the armed forces of several nations and put into service for assessment of battle dress uniforms, protective ensembles, and other combat-related clothing and equipment are publicly accessible. Two military systems with the most detailed available information are the Uncle Wiggly and Paul mannequin systems, both housed at U.S. Army facilities. Information on similarly functional U.K. and Canadian systems are discussed in Chapter 1.

Uncle Wiggly: Based at the U.S. Army Research Institute of Environmental Medicine (USARIEM) in Natick, Massachusetts, Uncle Wiggly was originally developed in 1984 with a copper-plated shell, cast-aluminum joints, and assorted heaters and sensors (Figure 2.1). Measurement Technology Northwest (in Seattle, Washington) updated the mannequin in 2004 with a thermal-control system consisting of signal conditioning, heater drivers and computer software, and a computer-controlled sweating system.

Uncle Wiggly features 19 individual heating zones, each with its own replaceable plug-and-play microcontroller to oversee temperature and fluid control and measurement. Software controls provide automatic steady-state detection; the operator can also program a work-cycle simulation and view instantaneous bar graphs and time-line graphs of any variable. Data updates are shown every second with calculations for number of watts of heat removed. Sweating occurs automatically through a series of valves and hoses that pump water through dozens of "weep holes" drilled through the metal that allow even and adjustable water distribution along the 19 sections. Motion capability includes the ability to swing arms and legs to simulate walking at speeds up to 3 mph. Testing is conducted in its own chamber that is controlled for temperature, humidity, and wind speed provided by a fan.

Paul: Paul (Figure 2.2) is a nonthermal, animatronic mannequin at the U.S. Army Soldier and Biological Chemical Command (SBCCOM) in Natick,

---

[1]Holmér, I. 2004. Thermal Manikin History and Applications. *European Journal of Applied Physiology* 92(6):614-618; Holmér, I., and H. Nilsson.1995. Heated Manikins as a Tool for Evaluating Clothing. *Ann. Occup. Hyg.*, Oxford University Press 39(6):809-818; Endrusick, T. L., L. A. Stroschein, and R. R. Gonzalez. Thermal Manikin History: United States Military Use of Thermal Manikins in Protective Clothing Research. Measurement Technology Northwest, *http://www.mtnw-usa.com/thermalsystems/history.html*. Accessed August 9, 2007.

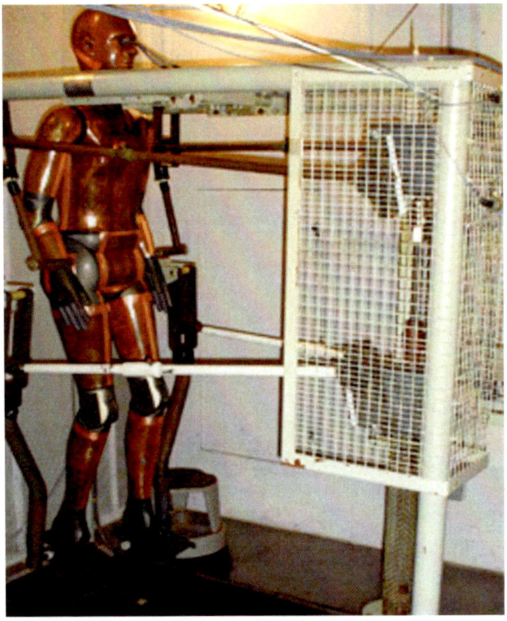

FIGURE 2.1 U.S. Army Research Institute of Environmental Medicine's (USARIEM's) Uncle Wiggly thermal mannequin in its own climate-controlled test chamber for measuring thermal and vapor resistance values of clothing ensembles.
SOURCE: "Modernized manikin: Uncle Wiggly resumes thermal testing after major 'organ' replacement." *The Warrior*, March-April 2004. *http://www.natick.army.mil/about/pao/pubs/warrior/04/marapr/index.htm*. Accessed August 7, 2007.

Massachusetts, but has been used for testing protective garments in a MIST chamber.[2] Built in 2002 by Creative Engineering, Inc. (in Orlando, Florida). Paul can be programmed for at least 20 motions.

Nonmilitary Systems

Commercial and academic groups have expended considerable time, effort, and expense to developing increasingly sophisticated and specialized thermal mannequins.[3] Some of the more prominent and innovative manne-

[2]U.S. Army Soldier Systems Center, Vapor Chamber Tests Chem/Bio Protective Prototypes. *http://www.amc.army.mil/amc/pa/Aug02Issue.html*. Accessed June 13, 2007.

[3]Nilsson, H. O. (2004). Comfort Climate Evaluation with Thermal Manikin Methods and Computer Simulation Models. *http://urn.kb.se/resolve?urn=urn:nbn:se:kth:diva-3726*. Accessed August 8, 2007.

FIGURE 2.2 U.S. Army Soldier and Biological Chemical Command's Paul mannequin cycling through motions in the Man-in-Simulant Test chamber.
SOURCE: "Vapor chamber tests chem/bio protective prototypes" (photo by Curt Biberdorf), *http://www.amc.army.mil/amc/pa/Aug02Issue.html*. Accessed August 7, 2007.

quin models are reviewed below. These are loosely categorized into "nonintegrated" systems that use discrete simulations of physiologic functions and "integrated" systems that are driven by human-physiology models.

Nonintegrated Systems

These systems use discrete simulations of physiologic functions.

Coppelius and variants: This line of mannequins originated in the VTT Technical Research Centre of Finland Laboratory of Plastics and Fiber Technology. Coppelius (Figure 2.3) features prosthetic joints for externally induced movement and different postures and computer-controlled heating

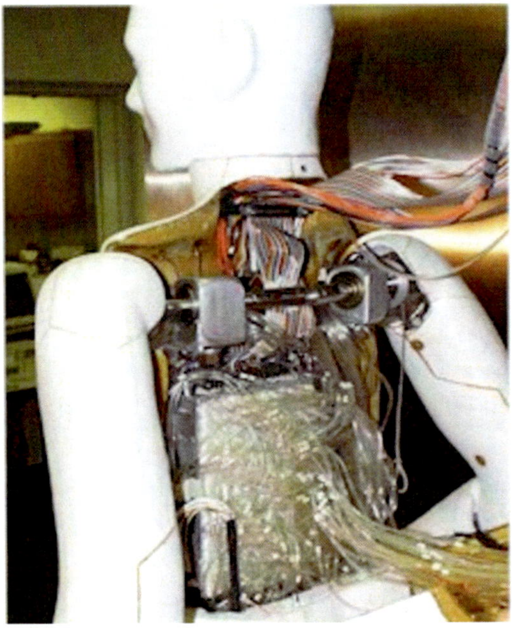

FIGURE 2.3 Coppelius mannequin.
SOURCE: Used with permission of North Carolina State University.

and sweating systems, with 18 individually controlled body sections and 187 sweating glands (distributed over the whole body except the head, hands, and feet). A Coppelius type of mannequin has been developed at the Center for Research on Textile Protection and Comfort of North Carolina State University through technology exchange with the Finnish VTT group.[4]

TOM III and SAM: Toyobo Corporation of Japan has manufactured two types of thermal sweating mannequins, TOM III (1980s) and SAM (1990s). They are differentiated primarily by their sweating mechanisms. TOM III uses water vapor from 220,000 "pores," whereas SAM sweats liquid water intermittently from 168 sites. SAM measures a broader array of variables (heat dissipation, "skin" temperature, and so on, in addition to temperature and humidity under clothing) and has more sophisticated computer control than TOM III.

---

[4]Additional information is available at *http://www.tx.ncsu.edu/tpacc/comfort/sweating_manikin.html*.

Walter: A team at the Institute of Textiles and Clothing of Hong Kong Polytechnic University, in Kowloon, Hong Kong, is using fabric technologies for its thermal sweating mannequin Walter (Figure 2.4). A waterproof but moisture-permeable fabric "skin" covers 1.5 m^2 of surface area and simulates the evaporation of sweat by moisture transfer of internally circulated body-temperature water from the mannequin core through tiny pores. The "skin" can be unzipped and interchanged with different versions to simulate different rates of perspiration.

Newton and Huey: Measurement Technology Northwest (MTNW), in Seattle, Washington, which developed the most recent version of USARIEM's Uncle Wiggly system, also produces several commercial systems for simulation of human physiologic temperatures and sweating. Its Newton and Huey models are constructed of aluminum-filled fiberglass-epoxy and aluminum, respectively; articulated joints and external frame allow for movement. Customizable and independent thermal zones—14-45 for Newton

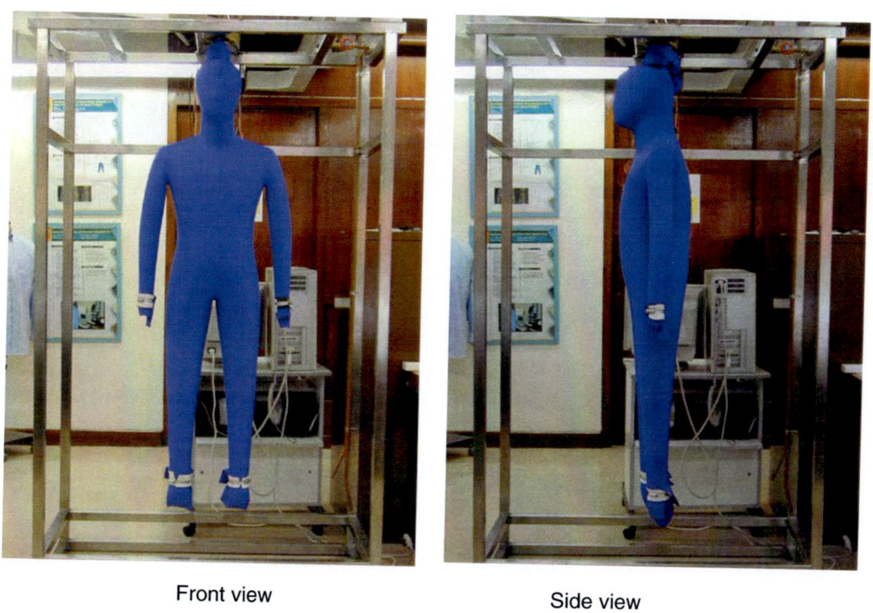

FIGURE 2.4 Sweating mannequin Walter.
SOURCE: Fan, J., Y. Chen, and W. Zhang. "A Perspiring Fabric Thermal Manikin: Its Development and Use." *Proceedings of the Fourth International Meeting on Thermal Manikins*, EMPA Switzerland, Sept. 27-28, 2001.

and up to 30 for Huey—can be used with an optional removable fabric sweating skin with an output capacity of 50-1,000 mL/h-m^2.

JUN: Built at Bunka Women's University, in Tokyo, with assistance from the Japan Society for the Promotion of Science, JUN incorporates several of the capabilities and features of previously discussed mannequin systems, namely, independently controlled and zoned thermal and sweating systems, fabric "skin," and externally inducible movement. Separation of the core and a 17-segment external shell allows distinct simulation of core and surface temperatures.

Integrated Systems

Integrated systems are driven by human-physiology models.

Advanced Automotive Manikin (ADAM): Actively used for research at the National Renewable Energy Laboratory (NREL), ADAM is another mannequin assembled by MTNW (Figure 2.5).[5] Measuring 61 kg and 175 cm and designed primarily as an automobile occupant, ADAM has 126 individually controlled stand-alone surface zones, each with a surface area of 120 cm^2 and integrated heating, temperature-sensing, sweat distribution and dispensing, and a heat-flux gauge and a local controller to manage closed-loop operation. Zone skin temperature is determined by an array of thermistors, and the sweating surface is all metal and optimized for thermal uniformity and response speed. Novel features include wireless control, a 24-V battery pack for two hours of autonomous continued operation, and a respiratory-simulation system for physiologic ventilation and heating with humidification of ambient air at up to 8 L/min.

The most unusual aspect of ADAM is its numeric physiologic model.[6,7] As temperatures are manipulated in ADAM's environment, the resulting skin heat-transfer rates are reported to a physiologic computer that uses mannequin conditions to generate prescribed and appropriate skin temperatures, surface sweat rates, and breathing rates. A loop feedback provides ever-changing measurements for assessment of human thermal comfort in

[5]Advanced Thermal Manikin. Measurement Technology Northwest. *http://www.mtnw-usa.com/pdf/ADAM.manikin.specs.pdf.*

[6]Paul, H., L. Trevino, G. Bue, J. Rugh, R. Farrington, and C. King. Phase II Testing of Liquid Cooling Garments Using a Sweating Manikin, Controlled by a Human Physiological Model. Doc 2006-01-2239. SAE International.

[7]Farrington, R., J. Rugh, D. Bharathan, H. Paul, G. Bue, and L. Trevino. Using a Sweating Manikin, Controlled by a Human Physiological Model, to Evaluate Liquid Cooling Garments. Doc 2005-01-2971. SAE International.

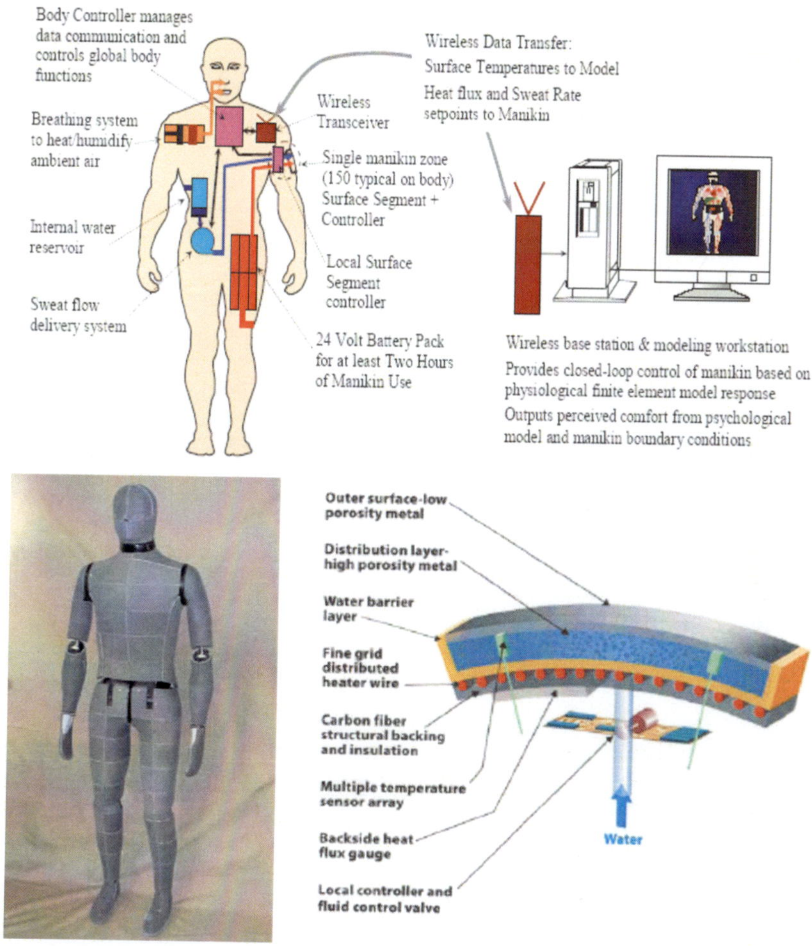

FIGURE 2.5 ADAM, the advanced thermal manikin.
SOURCE: Burke, R., and R. McGuffin, "Development of an Advanced Thermal Manikin for Vehicle Climate Evaluation" *Proceedings of the Fourth International Meeting on Thermal Manikins*, EMPA Switzerland, Sept. 27-28, 2001; R. Burke, Measurement Technology NW.

a dynamic environment. A separate thermal-comfort model predicts human perceptions of comfort under environmental conditions.

Sweating Agile Thermal Manikin (SAM): The Swiss Federal Laboratories for Materials Testing and Research (St. Gallen, Switzerland) has developed SAM, which is equipped with 26 shell parts and 4 heated joints in 30 separately heated sectors constructed of plastic mixed with aluminum powder (Figure 2.6). Electronics and software enable each sector to be heated to a constant temperature or with constant power using just one power supply for all shell parts. The maximal total heating power is 1.2 kW, the equivalent of very high-level human muscular activity.

One hundred twenty-five sweat outlets are distributed over the mannequin surface and positioned to ensure roughly human sweat distribution. Special pads cover the outlets to ensure that all water evaporates at low

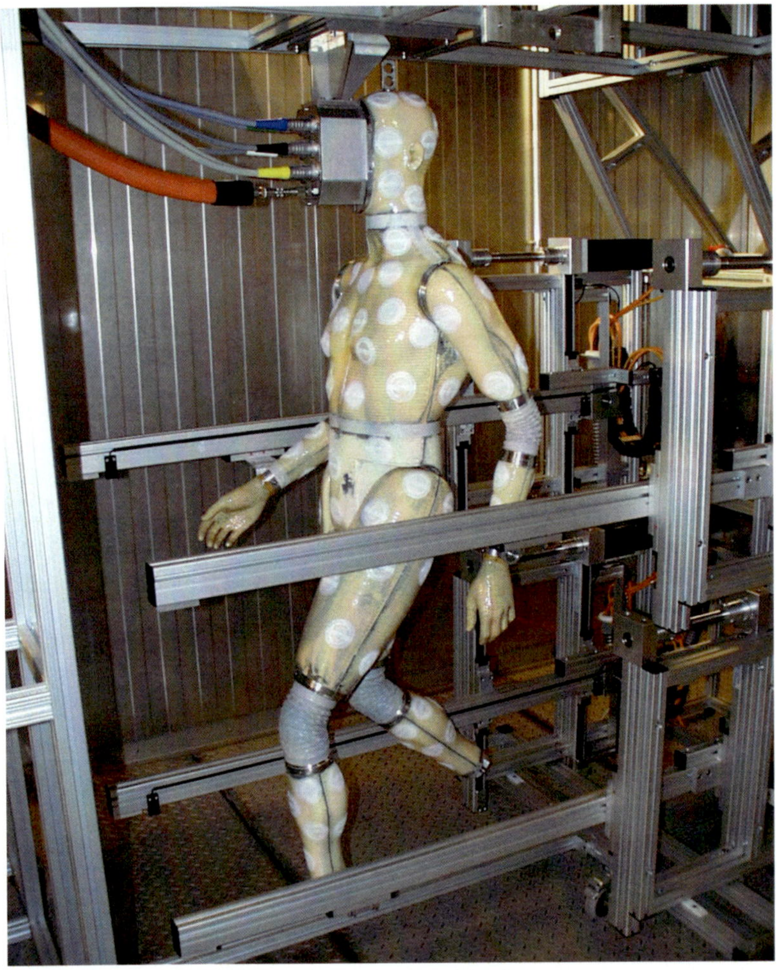

FIGURE 2.6 Sweating Agile Thermal Manikin (SAM).
SOURCE: SAM Sweating Agile Thermal Manikin of EMPA, Switzerland.

sweat rates to simulate insensible sweating or as both vapor and liquid water at higher sweat rates to simulate sensible sweating. Total sweat rate is controlled with a precision balance to measure the reduction in supply-tank water outside SAM; total moisture within clothing is determined by monitoring SAM's weight. The sweat rate can be varied from 20 mL/h to at least 4 L/h. Heating power, sweat rate, and body movements are linked for active-exercise phases to simulate human activities.

Joints at the shoulders, elbows, hips, and knees enable each limb to be moved in a vertical plane. Each limb is connected to a two-axis linear drive mechanism with movement curves defined as series of points with spline interpolation to ensure smooth curving, acceleration, and deceleration. Repetitive body movements, such as walking and climbing, can be performed during active testing by using a specially developed supporting frame. SAM is positioned in an environmental chamber that can be operated at temperatures of –30°C to 40°C and relative humidity of 30-95 percent. High wind speeds can be simulated by a wind generator positioned in front of SAM. The control unit is positioned on a trolley outside the chamber; heating and sweating supplies are connected through the face.

Other Temperature- and Perspiration-Related Systems

Several mannequin systems with environmental characteristics and capabilities not covered by the PETMAN specifications are also available.

The PyroMan Thermal Protective Clothing Analysis System is an adult-size flame-resistant mannequin system with 122 heat sensors distributed uniformly over the body (excluding hands and feet).[8] Each of PyroMan's sensors covers 0.82 percent of body area and is individually calibrated to ensure accurate reading of temperature and calculation of surface heat flux. The temperature readings, in conjunction with a one-dimensional transient heat-conduction model, are used to determine the heat flux experienced at the sensor's surface as a function of time.

A liquid-integrity testing system has also been described. A liquid-absorbing inner garment is placed over a mannequin and underneath the garment being tested to show penetration by test liquid. Water is the principal challenge agent; it is often treated with a surfactant to increase test severity. The testing duration is usually extended so that clothing design problems or defects will show up more readily. Test results are reported as pass or

---

[8]Barker, R. L. *A Review of Gaps and Limitations in Test Methods for First Responder Protective Clothing and Equipment*. January 31, 2005. Prepared for the National Institute for Occupational Safety and Health available at *http://www.cdc.gov/niosh/npptl/pdfs/ProtClothEquipReview.pdf*. Last accessed August 7, 2007.

fail, depending on the detection of liquid marks on the liquid-absorbing inner garment.

Cooling Systems

Cooling systems separate from the heating and sweating systems described above may be necessary for the combination of PETMAN functions. No sophisticated cooling systems for mannequins are available. Current humanoid and android robots, such as ASIMO and Hubo (see Chapter 4), normally use passive forms of heat removal; internal fans to circulate air and exhaust heat have been used infrequently. These robots' internal temperatures have been described as "warm to touch" and "hot to touch" with sustained activity (about 35-50°C; personal communication, Dr. J. H. Oh, Korea Advanced Institute of Science and Technology).

Cooling systems for human use have been developed for space exploration. The needs of astronauts on extravehicular missions have resulted in cooling garments that assist in thermoregulation. Available liquid cooling and ventilation garments use water at constant flow with vent ducting to remove more than 2000 BTU/h. Sublimator cooling, thermoelectric cooling, and cryogen-based heat-exchange systems are being actively investigated in combination with liquid cooling garments. Descriptions of several feasible technologies applicable to PETMAN have been provided by Grant Bue (National Aeronautics and Space Administration Johnson Space Center), including advanced heat pumps, a mini vapor compressor, "super ice," and vortex tubes.

Respiration and Ventilation

Respiratory physiology has been simulated for different purposes in different ways at various levels of verisimilitude. Mathematical models are available for numeric modeling of gas-exchange characteristics, and physical "lungs" have been replicated in isolation and in mannequins for simulation of mechanical aspects of ventilatory function.

Mechanical Lung Systems

- Physical lung-like systems have been developed for a variety of applications. Mechanical-ventilator training needs and development and testing requirements have resulted in isolated "lungs" that accurately simulate physicomechanical lung properties, such as airway resistance, lung volume, and compliance. A low-cost, fully mechanical multilobar lung from the University of Canterbury's Department of Mechanical Engineering (Christchurch, New Zea-

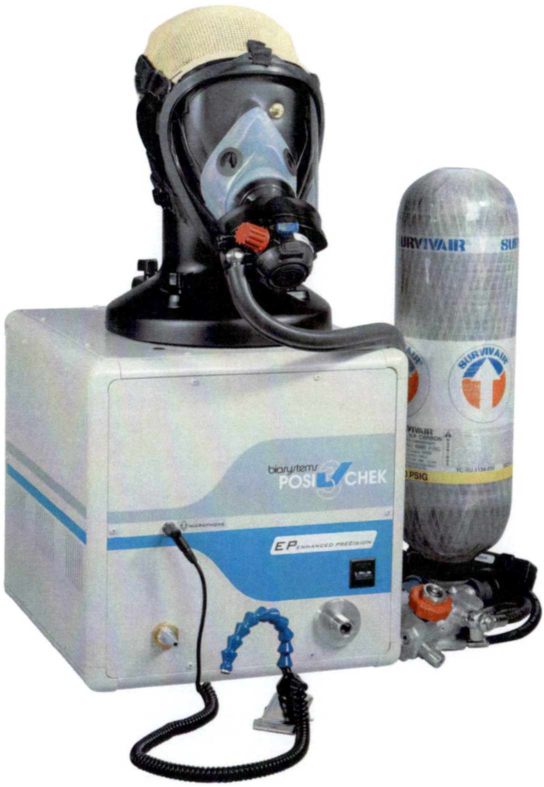

FIGURE 2.7 PosiChek[3] air-supplied breathing apparatus tester.
SOURCE: Used with permission of Biosystems LLC, Middletown, Connecticut.

land) allows accurate replication of several aspects of a passively breathing lung.[9]

- Equipment designed to test self-contained breathing apparatus (SCBA) and gas-mask assemblies are commercially available. Using large pistons or cams, these devices generate substantial airflow and replicate human ventilation to assess mask utility (see Figure 2.7).

[9]Chase, J. G., T. Yuta, K. J. Mulligan, G. M. Shaw, and B. Horn. 2006. A Novel Mechanical Lung Model of Pulmonary Diseases to Assist with Teaching and Training. *BMC Pulmonary Medicine* 6:21.

Mannequin Lung Systems

Technologic advances have generated important opportunities for enhanced medical education and training in the last two decades. Improved instructional devices are now being used in dedicated environments worldwide to teach safe patient care. Mannequin-based patient simulation is one such instrument, and it has applicability to PETMAN respiration requirements.

Human Patient Simulator: Commercialized as a mannequin system with the ability to simulate human patients undergoing anesthesia, the Human Patient Simulator (HPS) from Medical Educational Technologies, Inc. (METI), in Sarasota, Florida, features interconnected cardiovascular, pulmonary, pharmacologic, and metabolic computer models and bellows-driven lungs. Powered externally by a gas-driven pneumatic variable-pressure and variable-volume rack system, the lungs are capable of actively metabolizing inspired gases and anesthetic agents and exhaling the expected by-products of cellular metabolism, such as CO_2.

SimMan and SimBaby: These mannequins feature simple lungs and lack physiologic models. SimMan and SimBaby are manufactured by Laerdal Medical Corp., Wappingers Falls, New York. Costing substantially less than the METI HPS system, these mannequins have ventilatory functions that are powered by external compressors and gases.

NeuroDimension Inc. A bellows-less lung system has been developed by personnel at NeuroDimension Incorporated, and the College of Medicine Biomedical Engineering, and McKnight Brain Institute of the University of Florida at Gainesville.[10] Using a fixed-volume pressure controller to simulate spontaneous breathing, this model can simulate carinal pressure for simulation of actively breathing or ventilated patients and can simulate tidal volumes of 400 and 500 mL with flow rates of 4.3-5.7 L/min.

Physiology Integration

Integration of the different aspects of simulated human physiology as required by PETMAN has been accomplished to various degrees in the mannequin systems described above. Temperature and perspiration have been linked in thermal-mannequin programs to simulate metabolic states reflecting different levels of human activity; in the case of NREL's ADAM,

[10]Meka, V. V. and J. H. van Oostrom. 2006. Bellows-less Lung System for the Human Patient Simulator. *Medical and Biological Engineering and Computing* 42(3):413-418.

a rudimentary ventilatory system has also been added. As for simulators in medical education, sophisticated mathematical models of human cardiopulmonary physiology have been engineered to drive accurate and precise lung systems housed in human-shaped mannequins. Documentation of attempts to form an interface between warm, sweating mannequin systems and systems that breathe realistically seems not to be available.

Available systems do not address the question of whether mannequin-wide thermoregulation is possible in the context of a multifunction, moving, sensing computer-controlled human-shaped robot. Although a static mannequin can be warmed and made to sweat appropriately, no commercially available system has the capacity to cool a moving robot's actuator systems and onboard computers while heating its surface "skin." The PETMAN requirements for dynamic motion control and balancing of the robot may interfere with thermal and sweating regulation, particularly if the latter must simulate exercise-related physiologic functions realistically (see chapters 4 and 6 for more discussion of these potential complications).

DESIGN CHALLENGES AND HOW THEY MAY BE ADDRESSED

The human-physiologic simulation aspects of the PETMAN mannequin functions will be restricted by the need to combine physiologic software and engineering requirements (power, space, supply, and exhaust). Those issues are discussed below in the context of relevant PETMAN threshold (T) and objective (O) requirements (provided in italics). Chapter 6 presents further discussion regarding system integration and physical incorporation of physiologic-simulation components into an ensemble-compatible human-shaped package that houses chemical-agent sensors and robotic elements and uses materials that are resistant to toxic chemical materials.

Temperature and Perspiration

To maintain a constant mannequin "skin" temperature, the PETMAN system may require heating, cooling, or both. Whereas thermal mannequins contain only mechanisms necessary to generate heat and perspiration, PETMAN will contain many systems, each of which may have substantial thermal output. Furthermore, the equivalent of a circulatory system may become mandatory under such circumstances if simultaneous heating and cooling are required in different body segments.

Regulated heat and sweat generation for simulation of human skin temperatures and perspiration is possible today with dedicated mannequin systems. Mannequin heating is accomplished through heating wires or coils embedded in mannequin body-segment surfaces and controlled by power distribution. Mannequin sweating mechanisms use transmission of water as liquid or vapor into conduits and pores with selectable flow rates.

Commercially available systems do not have control and management of excess heat in robots or mannequins, because most applications have not required precise thermoregulation. Depending on individual component characteristics of the PETMAN system, there will probably be a net excess of heat generated by various sources, in which case cooling may be required. Furthermore, the mannequin core, compartment, and surface may vary substantially in temperature, given the nonhomogeneous distribution of components, and will probably require differential heating and cooling. MTNW has described having some experience in temperature regulation through heat exchange between warmed segments and "sweat"-circulation systems that carry cool liquids (personal communication, Rick Burke, MTNW).

The following—electronic control, mechanical construction, and system integration—are some of the key design challenges with respect to temperature and perspiration:

Electronic control

- Method of linking temperature and perspiration with respiration, metabolism, and motion:
  - Fixed heating, sweating, and ventilation output rates (open loop).
  - Fixed heating, sweating, and ventilation target values (closed loop, level 1).
  - Variable heating, sweating, and ventilation output rates and target values (temperature and perspiration linked through metabolic model).
  - Zoned heating and sweating (closed loop, level 2).
- Thermoregulation (concurrent heating and cooling of different body regions).

Mechanical construction

- Source of heat (power) and sweat (liquid and gas).
- Source of cooling (power and coolant).
- Exhaust.
- Durability (such as susceptibility to freeze damage and heat-related damage).
- Tether or no tether?

System integration

- Space limitations.
- Control for heat generation by other systems (such as CPU, transmitters, and sensors).
- Compatibility with other systems (device signatures, sources, and exhaust).
- Construction (resistance to chemical agents and cleanability).

Size specifications will reduce the possibility of temperature simulation owing to physical restrictions on the layout of heating, sweating, and cooling mechanisms and their supply sources and distribution; power limitations; exhaust requirements; and requirements imposed by other systems in PETMAN. All these will make meeting the anthropometric requirements extremely challenging even in the case of a tethered system (see chapters 4 and 6 for more discussion).

In isolation, mannequin skin temperature can be regulated by body region with current technologies with or without activity-dependent and region-dependent variability. PETMAN mannequin challenges will be related to physical restrictions on the layout, supply sources, and distribution of heating and cooling mechanisms; heating and cooling power limitations; heat and coolant exhaust requirements; interference from other heat sources in the mannequin distribution of heating or cooling to achieve targeted regional mannequin-skin temperatures; integration of control, space, and material requirements with the entire PETMAN system; and temperature regulation under extreme test-chamber conditions.

If heating and cooling are required simultaneously in different mannequin parts, a "circulatory system" will probably be required to establish demand-regulated regional distribution of centrally controlled liquids (or gases) to achieve temperature set points. Heat-exchange systems, liquid nitrogen- or CO_2-based systems, and vortex-tube designs are possibilities but have not been developed in the context of humanoid mannequin assemblies.

In isolation, perspiration can be regulated by body region with or without activity- or region-dependent variability with current technologies. PETMAN mannequin challenges will be related to physical restrictions on layout of the sweating mechanism, supply source, and distribution; sweat pumping-power limitations; sweat exhaust requirements; liquid or vapor interference with other PETMAN systems (for example, the effect of under-ensemble moisture on agent-detection sensors; integration of control, space, and materials requirements with the entire PETMAN system; and simulating sweating under extreme test-chamber conditions.

The PETMAN mannequin requirement for perspiration has described water and water vapor as acceptable sweat equivalents (current sweating mannequins use only water or water vapor), so the feasibility analysis does not address the biochemical and physiologic replication of electrolytes, oils, and other constituents of human sweat and their accompanying chemical reactivity and biomechanical implications.

Current thermal sweating mannequins and human-physiologic simulation mannequins are capable of continuous recording of skin temperature, respiration rate, and perspiration rate in time in one second increments.

Respiration and Ventilation

Positive-pressure mechanical simulation of human respiration and ventilation with regulatory mechanisms is possible now with dedicated mannequin systems. Mannequin respiration and ventilation are accomplished through "lung" reservoirs in the mannequin thoracic cavity that are powered by external pneumatics or bellows. These lungs create chest-wall movement and simulated air or gas movement transmitted through either a nonanatomic pneumatic or an anatomic airway system. Depending on the mannequin system, gas exchange (for example, oxygen consumption and CO_2 exhalation) can also be simulated. Software controls range from simple respiratory rate and volume regulation to automated determination of pulmonary measures corresponding to whole-mannequin conditions, such as cardiac arrest or septic shock.

Size specifications will limit respiration and ventilation simulation because of design limitations imposed by whether a pneumatic or actuated chest-wall movement system is implemented (see below), power limitations, and interference with other PETMAN subsystems.

The PETMAN mannequin requirement for respiration and chest-wall movement has been described as predicated on the need to replicate under-ensemble conditions—such as compartment "hot spots," pressure differentials, and airflow—that may affect ensemble leakage or penetration and agent transport and distribution in the ensemble. The issue of respiratory system air flow at 50 L/min can be addressed either with or separately from chest-wall movement. Specifically, the requirements theoretically could be met at different levels (Figure 2.8) with a unified mannequin respiratory system or with discrete but linked component systems.

Unified System (Anatomically Correct)

A respiratory system based on human anatomy and depending on negative intrathoracic pressures will be one candidate for PETMAN airflow and chest-movement needs. Like human respiration and ventilation, this type of system would require a true self-contained intrathoracic bellows system

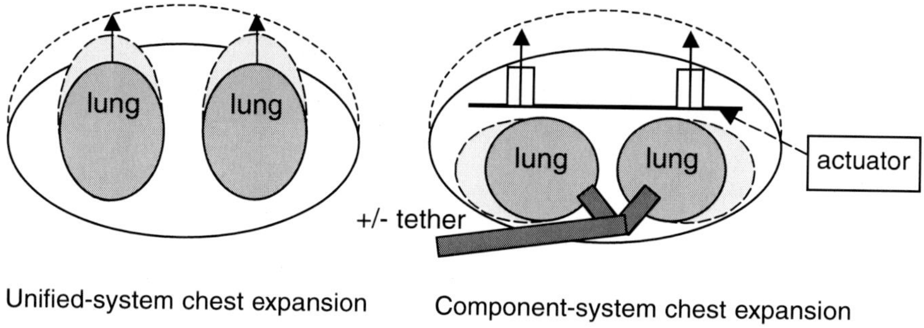

FIGURE 2.8 Comparison of systems that simulate human respiration and ventilation. Axial cross-sections of mannequin thoracic cavity are shown.

(diaphragm, compliant chest, and lungs) capable of inducing sufficient negative pressure to draw air from the environment into the chest cavity and then exhale it. Such a system, with internalized pneumatic systems to drive mannequin breathing, has not been reported in commercially available mannequins but could be made available with short-term research and development.

Component Systems (Nonanatomic)

If movement of air through the mannequin's nose, mouth, airway, and lungs is handled separately from chest-wall movement, two nonanatomic mechanisms may be used to achieve each objective independently: airflow and chest movement.

- Current METI mannequins use an extrathoracic (tether-connected) bellows system to generate both negative pressure to draw air in and positive pressure to exhale. Their bellows mechanism works with a simple airway (basic but accurate facial features, oropharyngeal and nasopharyngeal structures, dentition, tracheal and bronchial structures, and bag-reservoir lungs) for adequate simulation of human respiratory and ventilatory airflow. Less critical breathing-related biophysical properties—such as airway resistance, dead-space ventilation, and positive end-expiratory pressure (PEEP)—could be replicated with refinements of mannequin anatomy. (Alveolar metabolism of inhaled materials, exhalation of human gaseous metabolic products, and mask-seal and airway or alveolar absorption of agent will not be addressed in this analysis.)

  Separate from bellows mechanisms, a piston- or cam-based

system (as used in respiratory-mask testing systems manufactured by ATI and Biosystems) may generate negative inspiratory forces sufficient for PETMAN respiration and ventilation simulation. Published specifications describe flow up to 102 L/min through a standard SCBA mask assembly. Space and power requirements will probably be greater than for bellows systems because of the mechanisms and higher pressures and flow volumes generated.

Another theoretical alternative is to interpose a two-way jet ventilator system in the mouth of the mannequin; the inward-facing and outward-facing jets would simulate inspiration and exhalation, respectively. Disadvantages include nonphysiologic airflow patterns and outlet pressures.

- Mannequin chest expansion can be independently accomplished with intrathoracic actuators (see Figure 2.8). METI has announced a self-contained, untethered battery-operated mannequin that creates chest-wall movement without lung pneumatics).

Current mannequins that simulate human physiology are capable of continuous recording of skin temperature, respiration rate, and perspiration rate in one second increments.

Physiology Integration

Chapter 6 discusses PETMAN system integration in detail. This section is limited to the integration of human-physiology simulation into the entirety of PETMAN. Mannequin breathing, surface temperature, and perspiration may be integrated at several levels: full dynamic, limited dynamic, or preset integration.

Full Dynamic Integration: Comprehensive linking and feedback control of all PETMAN systems with human-physiology simulation would be necessary for a mannequin capable of detecting chemical agents and responding in a manner that resembles human pathophysiology, for example, a reduction in heart rate caused by sarin exposure. This level of integration will be difficult to achieve.

Limited Dynamic Integration: PETMAN human-physiology simulation may be designed to operate discretely from sensor systems but able to self-regulate to achieve target vitals signs and perspiration rate. Physiologic models accomplishing this have been described by NREL and MTNW, whose ADAM mannequin is not complicated by additional onboard systems that generate heat and drain power. Use of a "metabolism" model to drive physiologic simulation is a potential solution (see Figure 2.9).

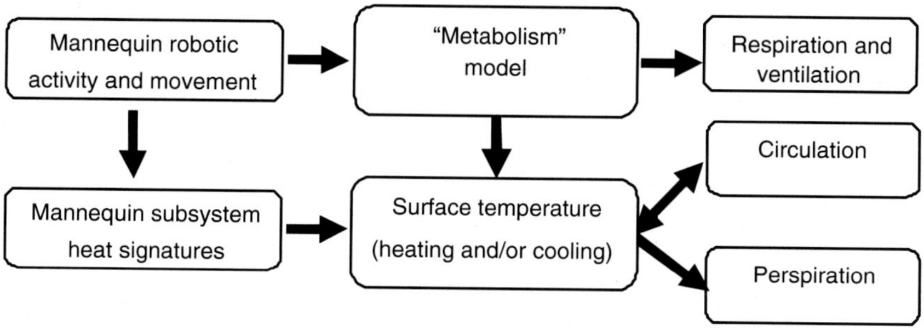

FIGURE 2.9 Diagram of potential mannequin "metabolism" model.

Preset Integration (Pseudophysiology): The least complicated method of linking the different human-physiology simulation components is to define preset states (such as "resting," "walking," "running," and "crawling") and their corresponding physiologic characteristics. Setting respiratory rate, tidal volume, surface temperature, and sweating rate for a defined activity simplifies feedback to achieve target values.

Fitting the component subsystems necessary only for human-like thermoregulation, sweating, respiration, and ventilation should be possible in a mannequin sized as a 50th percentile male. For example, iStan is completely self-contained, has skin made of thermoplastic elastomer, has a realistic weight, and uses pressurized water to create sweating. An area is also available in the belly to allow for additional simulation capabilities.[11]

However, as discussed in more detail in Chapter 6, successful positioning and complete packaging of the physiologic simulation subsystems alongside the sensor, robotics, and power systems needed to meet all PETMAN requirements are likely not feasible with current technologies without a tether. In any size state-of-the art robots, actuators fill the volume of the arms, legs, shoulders and hips, but space is typically available in the torso and head. In a 50th percentile male-sized PETMAN that meets all the objective requirements, about 3 L in the head and about 20 L in the torso would be available for all the other capabilities. The water for respiration, sweating, and cooling could be stored in the head. The torso space would be available—but insufficient—for all the remaining computer, electronic, heating, sensing, and battery components. Approximately 24 L of the mannequin would be needed to house just the batteries to power PETMAN for 12 hours.

[11]Presented by Carlos Moreno, Medical Education Technologies, Inc.

On the basis of the extensive physical-activity requirements specified for PETMAN, the onboard processing and robotic actuation (whether pneumatic, electric, or hydraulic) involved will require substantial energy stores and generate considerable heat and exhaust. Although a breathing-mannequin system at rest could potentially be made to thermoregulate and perspire appropriately in all its body segments, the technology to control mannequin heating and cooling dynamically to achieve the same homeostasis within the limits of operational testing is not now available.

COST-BENEFIT ANALYSIS AND TRADE-OFFS

The PETMAN requirements related to human physiology are analyzed in Table 2.1. The analysis includes discussion of the technical benefit, cost impact (disadvantages), near-term availability, and feasibility.

FEASIBILITY AND POTENTIAL ALTERNATIVES

Based on the information presented earlier in this chapter, it is feasible to simulate the independent components of physiologic conditions under the IPE (requirement 3.3.4, Appendix B) at least the threshold levels. However, given the space available in the mannequin (50th percentile male anthropometrics) and the heating, cooling, and sweating system requirements (power, supply, and exhaust demands), it is not feasible with current technologies to meet the human physiologic requirements with additional chemical sensing and robotics operational requirements in an autonomous nontethered PETMAN system.

Conclusion 2-1a: Simulation of chest-wall movement associated with respiration is feasible.

Conclusion 2-1b: Simulation of human respiration and ventilation with chest movement and respiratory tract airflow through external (tether-connected) pneumatics is feasible.

Conclusion 2-1c: Internalization of all systems necessary to generate chest-wall movements and simulated respiration and ventilation in the mannequin will require additional research and development.

Recommendation 2-1: The role of simulated respiratory tract airflow in the mannequin during IPE testing should be further clarified. Distinct from simulation of respiratory chest-wall movement for re-creation of under-ensemble airflow, the generation of accurate mannequin inspiratory and expiratory airflow through a mask respirator will require additional development and may duplicate current protocols for testing of IPE mask respirators in isolation.

Conclusion 2-2a: Simulation of human activity-dependent skin temperatures and sweating (with water as a sweat equivalent) should be feasible when space, power, electric, and exhaust requirements of other PETMAN subsystems are housed outside the mannequin via a tether.

Conclusion 2-2b: Integration of activity-dependent human-like skin temperatures and sweating into mannequin wide thermoregulation with differential heating and cooling through a circulatory system has not been achieved. That is, a static mannequin can be warmed and made to sweat appropriately, but no commercially available system has the capacity to cool a moving robot's actuator systems and onboard computers while heating its surface "skin." However, based on the currently available systems, this requirement may be feasible with additional short-term research and development.

Recommendation 2-2: Inasmuch as mannequin-wide thermoregulation is by definition an all-or-none proposition, its feasibility analysis does not lead to recommendations regarding potential alternatives or sacrifices in performance requirements.

Conclusion 2-3a: Mannequin respiration, skin temperature, and sweating should be controlled through either preset physiologic characteristics for defined IPE test activity states or a mannequin metabolism model designed to operate without physiologic linkage to sensor systems.

Conclusion 2-3b: Integration and installation of the necessary systems to achieve all human-physiology simulation requirements into a PETMAN that meets all specified operational requirements will encounter substantial challenges with the use of technologies that are currently available or are expected to be available in the near future. The presence of a tether may relieve some—not all—obstacles to this objective.

Recommendation 2-3: A phased development project with progression through transitional stages may alleviate several technologic hurdles that need to be overcome if all human-physiology simulation requirements are to be met in a robotic sensing PETMAN.

The above conclusions are in accordance with the feasibility analyses of other PETMAN system considerations, such as robotics (see Chapter 4) and systems integration (see Chapter 6).

TABLE 2.2 Cost-Benefit Analysis of Simulating Human Physiologic Characteristics as Part of PETMAN System

Requirement Number	Requirement Description	Technical Benefit
3.3.1	Tethered (T)	Easier, cheaper solution to power, thermal, mechanical issues
	Untethered (O)	No modification of IPE; potentially more realistic test challenge
3.3.3	Anthropometric requirements of 50th percentile male	Compatibility with protective ensemble
3.3.4.1	Fixed skin-temperature simulation by body region (T)	Intended to be easier to achieve than variable temperature based on region and activity

Cost Impact/ Disadvantages	Comments	Near-term Availability	Feasibility[a]
Requires nonstandard modification of IPE	Assume not just a simple tether: mechanical, power, coolant interface	Would still require integrating subsystems but appears feasible in 1-2 years	3
Power, thermal, mechanical challenges; probably insurmountable in required timeframe	—	—	1
Space limitations leading to restrictions on every system	—	—	1
Maintaining skin at controlled temperature, whether fixed or model-based, requires both heating and cooling regulation	Not much easier than variable skin temperature; challenge is in system integration	Heating subsystems are available from several mannequin manufacturers; cooling methods could be adapted from spacesuit technology	3

continued

TABLE 2.1 Continued

Requirement Number	Requirement Description	Technical Benefit
	Variable skin-temperature simulation by body region and level of physical activity or exertion (O)	More accurate simulation of temperature in IPE, hence realistic simulation of IPE internal and external environmental challenges
3.3.4.2	Fixed perspiration rate (T)	Intended to be easier or cheaper to achieve than variable perspiration rate
	Variability in perspiration rate based on level of physical activity or exertion (O)	Accurate simulation of moisture environment in IPE, capturing moisture effect on agent behavior
3.3.4.3	Fixed respiration rate: (a) combined	Small increase in accuracy of simulation of agent transport in IPE and in IPE mask

Cost Impact/ Disadvantages	Comments	Near-term Availability	Feasibility[a]
Maintaining skin at controlled temperature, whether fixed or model-based, requires both heating and cooling regulation	Not much more difficult than fixed skin temperature; challenge is in system integration	Heating subsystem are available from several mannequin manufacturers; cooling methods could be adapted from spacesuit technology	3
Complex liquid-delivery system available from several manufacturers	Not much less difficult than variable perspiration rate; challenge is in system integration	Subsystems are available from several mannequin manufacturers	3
—	Not much more difficult than fixed perspiration rate; challenge is in system integration	Subsystems are available from several mannequin manufacturers	3
Pneumatic or actuated system that simulates chest motion while driving respiratory flow	Current mannequin systems use positive pressure, unlike humans	Potentially applicable subsystem also available from respiratory-mask testing companies	3

continued

TABLE 2.1 Continued

Requirement Number	Requirement Description	Technical Benefit
	Fixed respiration rate: (b) respiratory flow only	Accuracy of simulation of agent transport in IPE mask
	Fixed respiration rate: (c) chest motion only	Accuracy of simulation of agent transport in IPE, that is, under-ensemble
	Variable respiration rate based on level of physical activity or exertion (O)	Additional small increase in accuracy of simulation of agent transport in IPE and in IPE mask
3.3.11	Decontaminate with no effect on next iteration of test (T)	Necessary to process multiple tests
	Decontaminate leaving negligible agent residual defined by DA PAM 385-61 (O)	—

Cost Impact/ Disadvantages	Comments	Near-term Availability	Feasibility[a]
Pneumatic or actuated system that drives respiratory flow	Positive or negative pressure not specified in requirements	Potentially applicable subsystem also available from respiratory-mask testing companies	2
Pneumatic or actuated system that simulates chest motion	May be insignificant compared with limb-motion effects	Used by some current mannequins	1
Preprogrammed control, variable-speed motor	—	Potentially applicable subsystem also available from respiratory-mask testing companies	1
Will probably require time, effort, resources between tests	—	—	1
Agent residue may complicate some sensor readings	—	—	1

continued

TABLE 2.1 Continued

Requirement Number	Requirement Description	Technical Benefit
3.3.13	System measure recording	Test-data logging for analysis
Nonrequirement issues: System integration	Physiologic integration	Necessary for all requirements to be met simultaneously
	Accurate modeling of skin, anatomy to represent mask seals	—
	Packaging	—
	Cooling	—

[a]3 = high feasibility, 2 = medium feasibility (achievable with substantial effort), 1 = low feasibility (extremely difficult to achieve)

Cost Impact/ Disadvantages	Comments	Near-term Availability	Feasibility[a]
Additional power, storage requirements	—	Available from several mannequin manufacturers	3
—	Significant design and development risk—combined robotic-physiologic mannequins are not available, nor is approach evident	—	1
—	Not clear whether mask sealing has high priority; could be addressed with separate test	—	2
Physical connections, such as tubes and wires, must fit within volume, thermal constraints	Technical challenge with no precedent in current state of art	—	1
—	Overall integration of movement, physiology, thermal requirements is far greater challenge than separate elements	—	1

3

Design Challenge: Mannequin Under-Ensemble Sensing

This chapter mainly discusses the following two PETMAN design challenges:

3.2.4 The study will determine the feasibility of designing a PETMAN system with an integrated under-ensemble sampling system that will allow for the collection of agent breakthrough data in real time (1-second increments).

3.2.6 The study will determine the feasibility of designing a PETMAN system capable of operating in fixed environmental chamber conditions (T) and a range of environmental chamber conditions (O) as defined in 3.3.6.1-3.3.6.5.

The design, fit, or size of the protective garment will influence chemical-agent penetration. When worn, clothing is subjected to pressure differentials across the garment due to breathing, wind, or the bellows effect created by body movement, which may force chemical-warfare agent (CWA) or toxic industrial chemicals (TICs) vapor or aerosol through the clothing fabric and closures. Bending and moving stresses clothing, particularly over the joints. To determine penetration of chemical-protective ensembles by chemical vapor or aerosol, it is necessary to test the entire suit system, including seams, closures, and areas of transition with other protective equipment, that is, at the ankles, waist, wrists, and neck. In addition, aspects of the PETMAN tests such as the environmental conditions of the test chamber will also impact CWA penetration through the soldier IPE.

In addition to the two design challenges given above, consideration will also be given to the PETMAN design challenge concerning mannequin support, which also impacts the feasibility of CWA sampling technologies:

3.2.1 The study will determine the feasibility of designing a PETMAN system to be tethered (T) or free standing/self-contained (O). A tethered PETMAN system design must not compromise the integrity of the individual protection ensemble equipment being tested on the PETMAN system. If a tethered design is selected the design must also minimize the impact to the whole ensemble operation.

RELEVANT PETMAN REQUIREMENTS

The relevant PETMAN requirements (Box 3.1) and various technologies for the detection of CWAs for potential use in the PETMAN mannequin are discussed below.

Other Considerations

Some conditions in addition to the PETMAN system design challenges and requirements designated above should also apply to ensure proper functioning of the sensor technology in the context of all other PETMAN requirements.

The technology should be:

- Unaffected by perspiration (water vapor) beneath the garment ensemble
- Unaffected by any potential interferents from garment material or residual decontamination cleanser and chemicals off-gassing
- Packaged in a durable device that can withstand rough handling and repeated use
- Portable enough to be conveniently placed in different locations on the test mannequin and beneath the protective garment
- Unaffected by "false-positive" readings (for example, readings that result from residual volatile additives in the protective-garment material, decontamination chemicals and cleansers, or from moisture interference)
- Currently available in a usable product form
- Have minimal impact on internal airflow within the suit

BOX 3.1
Relevant PETMAN Requirements

3.3.2 The PETMAN system shall be capable of operation for twelve (12) hours prior to requiring operational maintenance, three (3) months prior to preventive maintenance and six (6) months prior to calibration. (T) The PETMAN system shall be capable of operation for twenty-four (24) hours prior to requiring operational maintenance, six (6) months prior to preventive maintenance and twelve (12) months prior to calibration. (O) Operational maintenance is defined as the required maintenance procedures to prepare the PETMAN system for each test trial, for example, filling a perspiration reservoir, changing agent samplers or decontamination before the next trial. Preventive maintenance is defined as a maintenance event performed prior to a failure in order to prevent its occurrence.

3.3.5 The PETMAN system shall be compatible with current under ensemble chemical breakthrough sampling technologies, procedures, and equipment as defined in Test Operations Procedure (TOP) 10-2-022, Chemical Vapor and Aerosol System-Level Testing of Chemical/Biological Protective Suits (T) and designed to enable integration with real-time (1-second increments) sampling technologies, procedures, and equipment (O). At a minimum, sampling locations shall be the same as those defined in TOP 10-2-022 (see Figure 3.1 and Table 3.1).

3.3.6 The PETMAN system operation shall not be affected by the following chamber environmental conditions.
3.3.6.1 Temperature: 90°F ± 2°F (T); –25°F to 125°F ± 1°F, measured every 5 minutes (O)
3.3.6.2 Relative Humidity: 80% ± 3% (T); 0-100% ± 1%, measured every 5 minutes (O)
3.3.6.3 Wind speed: 0-10 mph ± 10% (T); 0-161 mph ± 2 mph (O)
3.3.6.4 Pressure: 0.25 iwg chamber vacuum maintained ± 2%
3.3.6.5 Liquid and vapor chemical agents including all nerve and vesicant agents, as well as the chemical simulants, triethylphosphate and methyl salicylate.

3.3.12 The PETMAN system shall utilize as many common commercially available parts and/or components as possible.

3.3.13 The PETMAN system shall record the following system parameters over time: skin temperature, respiration rate, perspiration rate, and total mass (in nanograms) of chemical vapor that penetrates/permeates through the protective ensemble. The PETMAN system shall record the start and stop time of each motion in 1 second increments.

NOTE: T=threshold and O=objective

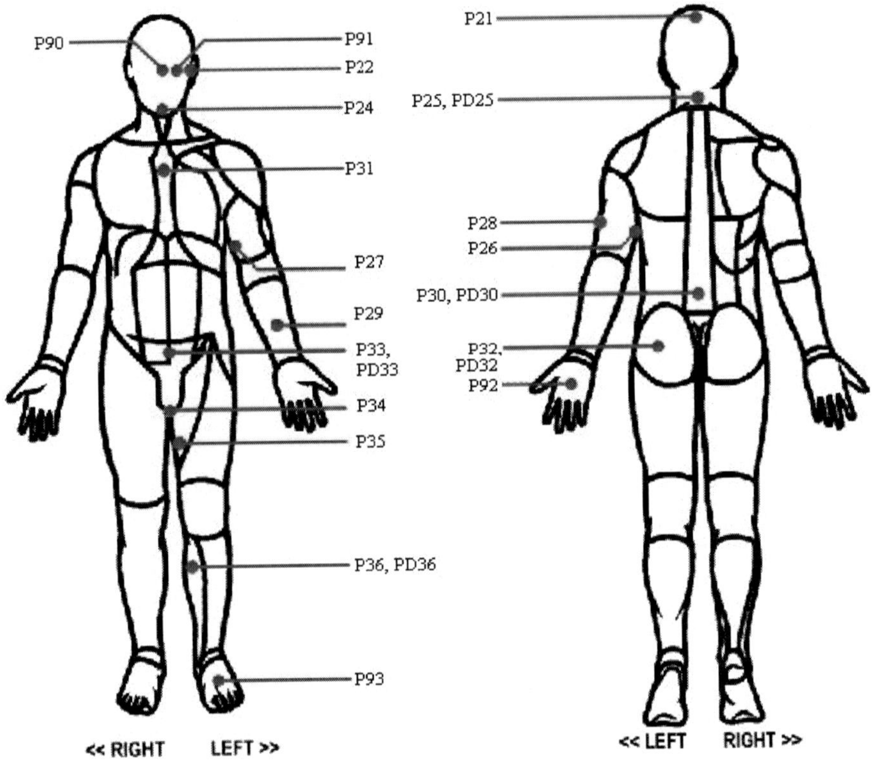

FIGURE 3.1 Sampling locations for the Passive Absorbent Devices (PADs) for Man-in-Simulant Test (MIST); Chemical Vapor and Aerosol System-Level testing of chemical/biological protective suits.
SOURCE: U.S. Army Developmental Test Command Test Operations Procedure (TOP) 10-2-022. December 14, 2005. Chemical Vapor and Aerosol System-Level Testing of Chemical/Biological Protective Suits. DTIC AD No. ADA440290.

CURRENT DETECTION TECHNOLOGIES

A chemical sensor uses chemical (and possibly biologic) reactions to detect and quantify a specific analyte. Such devices are usually self-contained and have a sorptive surface that interacts chemically with the analyte of interest (commonly a polymeric membrane or coating that contains a specific low-molecular-weight "doping agent"), a transducer that detects the chemical event occurring between the surface and the analyte, and supporting electronics or software that amplifies and reports the output signal

TABLE 3.1 Passive Absorptive Device (PAD) Placement Location Descriptions; Individual Protective Equipment for Selectively Permeable Membrane Materials Man-in-Simulant Test, as Shown in Figure 3.1

Position Number[a]	Description
P21	Scalp
P22	Left Ear
P24	Chin
P25, PD25[b]	Nape
P26	Armpit
P27	Inner Upper Arm
P28	Outer Upper Arm
P29	Forearm, Volar
P30	Mid Back
P31	Abdomen
P32, PD32[b]	Buttocks
P33, PD33[b]	Groin
P34	Crotch/Scrotum
P35	Inner Thigh
P36, PD36[b]	Inner Shin
P37	Blank Transport Sample for Quality Control (QC)
P38	Spike Transport Sample for QC
Don	Dress Area
Doff	Undress Area
P90	Nose Cup
P91[c]	Mask
P92	Glove (Hand)
P93	Boot (Foot)

[a]See Figure 3.1.
[b]Indicates duplicate PSD at these locations.
[c]PSD placed in transport case to monitor cross-contamination.

SOURCE: U.S. Army Developmental Test Command Test Operations Procedure (TOP) 10-2-022. 14 December 2005. Chemical Vapor and Aerosol System-Level Testing of Chemical/Biological Protective Suits. DTIC AD No. ADA440290.

from the transducer. The sorptive material is probably the most critical component of the sensor in terms of determining the sensitivity and selectivity, regardless of the transducer.[1]

[1]Grate, J. W., and D. A. Nelson. 2003. Sorptive Polymeric Materials and Photopatterned Films for Gas Phase Chemical Microsensors. *Proceedings of the IEEE* 91(6):881-889.

TABLE 3.2 Overview of Technologies Currently Available for Detection of CWAs, TICs, and TIMs

Technology	Principle
Chemicapacitor technology	An array of polymer-filled chemically sensitive capacitors, detects the presence of vapors in the air. A single 2 x 5-mm chip has the potential to contain an array of up to eight capacitors. A more detailed discussion is presented below.
Electrochemical sensors	These sensors operate by reacting with the gas or vapor of interest and producing an electric signal that is proportional to the gas or vapor concentration.
Fiber-optics-based sensor	A polymer-, porphyrin-, or enzyme-treated substrate is subjected to a specified range of light. The response of the detection system to the nonabsorbed radiation determines the types and amounts of chemical agents present. A more detailed discussion is presented below.
Flame-ionization detector (FID)	A FID is an ion detector that uses an air-hydrogen flame to produce ions. As components elute from the gas chromatograph's column, they pass through the flame and are burned, producing ions. The ions produce an electric current, which is the signal output of the detector. The greater the concentration of the component, the more ions are produced and the greater the current.

Many technologies exist for detecting CWAs, TICs and TIMs, and have been reviewed extensively elsewhere.[2] A list of some potential detection technologies for PETMAN is given in Table 3.2 and evaluated in Table 3.3. Three of these detection transducers at various stages of

[2]See the following reviews and others for more information on chemical and biological sensing technologies: Dadik, O., W. H. Land, Jr., and J. Wang. 2003. Targeting Chemical and Biological Warfare Agents at the Molecular Level. *Electroanalysis* 15(14):1149-1159; Seto, Y., M. Kanamori-Kataoka, K. Tsuge, I. Ohsawa, K. Matsushita, H. Sekiguchi, T. Itoi, K., Y. Sano, and S. Yamashiro. 2005. Sensing Technology for Chemical-Warfare Agents and Its Evaluation Using Authentic Agents. *Sensors and Actuators, B: Chemical* 108:193-197; Sun, Y., and K. Y. Ong. *Detection Technologies for Chemical Warfare Agents and Toxic Vapors*, CRC Press (2004).

TABLE 3.2 Continued

Technology	Principle
Ion-mobility spectrometer (IMS)	A radioactive source is used to ionize the target gas; the resulting ions are directed into a drift chamber where they are separated by mass and charge. The amount of ions of each mass or charge collected determines the amount and type of gas or vapor molecules present.
Passive adsorptive devices (PADs)	PADS technology uses packets of Tenax® that adsorb the agent simulant methyl salicylate. Afterward, the simulant is desorbed from the Tenax and quantified. This technique is currently used in the Man-in-Siumulant-Test.
Photoionization detector (PID)	A PID uses an ion detector having high-energy photons, typically in the UV range, to produce ions. As components elute from the gas-chromatograph column, they are bombarded by high-energy photons and are ionized. The ions produce an electric current, which is the signal output of the detector. The greater the concentration of the component, the more ions are produced and the greater the current.
Surface acoustic wave (SAW)	Detection is based on the change in the propagation of an acoustic wave over a piezoelectric substrate treated with a polymeric coating. A more detailed discussion is presented below.

development—chemicapacitor, fiber-optics, and surface acoustical wave (SAW)—were chosen as illustrative examples for evaluating the feasibility of the PETMAN requirements in more detail. The chemicapacitor technology is commercially available but not widely used, fiber-optics based is not commercially available but has the potential to be developed with the next year, and SAW based is commercially available and widely used in many application areas. In Tables 3.4 through 3.6 the three technologies and the current passive absorptive devices (PADs) currently used in the MIST are compared based on how they all address the PETMAN requirements.

Chemicapacitor Detection Technology

Chemicapacitor sensors measure the change in dielectric constant of selectively adsorbing materials such as a polymer or single-walled carbon

TABLE 3.3 Ratings of Detection Technology for Various Characteristics (L = little or no capability, M = medium capability, H = highest capability or most desirable)

	Chemicapacitor	Electro-chemical	Fiber-Optics	FID	IMS	PADs	PID	SAW
Range of CWAs detected	H	L	H	L	H	Unknown	M	H
Range of TICs detected	L	M	H	L	M	Unknown	M	L
Selectivity for agents detected	H	M	H	L	L	Unknown	L	H
Sensitivity to CWAs	H	L	M-H (depends on system)	L	H	Unknown	L	H
Relative size or footprint (smaller is more desirable)	M	L	H	L	L	H	L	M
Start-up time (shortest time is most desirable)	H	M	H	M	M	H	M	M
Response time (≤ 1 s is most desirable)	M	M	H	M	M	L	M	M
Potential for wireless power source	Yes	Yes	Yes	Yes	Yes	Not powered	Yes	Yes

TABLE 3.4 Comparison of Various Detection Technologies with PADs in Meeting Threshold PETMAN Program Requirements

Requirement	Chemicapacitor	Fiber Optics	PADs	SAW
Commercialized	Yes	No	No	Yes
Can be placed at all sampling locations defined in TOP 10-2-022 (see Section 3.3.5)	Yes	Yes	Yes	Yes
Capable of detecting specified agents—nerve agent GB and vesicant HD—and their simulants, triethyl phosphate and methyl salicylate, respectively (see Section 3.3.6)	Yes[a]	Yes	Yes	Yes
Capable of detecting nanogram levels of agent or simulant penetration or permeation	Yes	Yes	Not applicable	Yes
Capable of operating for 12 hours before requiring operational maintenance, three months before preventive maintenance, and six months before to calibration (see Section 3.3.2)	Yes[b]	Yes	Not applicable	Yes
Compatible with current under-ensemble chemical-breakthrough sampling technologies, procedures, and equipment as defined in TOP 10-2-022 (see Section 3.3.5)	Yes	Yes	Yes	Yes
Operational at (not affected by) a temperature of 90 ± 2°F (see Section 3.3.6.1)	Yes	Yes	Yes	Yes

[a]Limits of detection have to be determined or confirmed.
[b]Long-term testing has yet to be performed.

TABLE 3.5 Comparison of Various Detection Technologies to Meet Objective PETMAN Program Requirements

Requirement	Chemicapacitor	Fiber Optics	PADs	SAW
Powered in such a manner that it does not require tethering to a power source, for example, battery-powered and wireless (see Section 3.2.1)	Yes	Yes	Yes	Yes
Capable of operating for 24 h before requiring operational maintenance, 6 months before preventive maintenance, and 12 months before calibration (see Section 3.3.2)	Yes[a]	Yes	Not applicable	Yes
Able to detect and report agent or simulant permeation or penetration in real time in 1 second increments (see Section 3.3.5)	Yes	Yes	No	No
Operational at (not affected by) temperatures of –25 to 125 ± 1°F (see Section 3.3.6.1); note, however, that if a mannequin skin that evolves metabolic heat is developed, a temperature of –25°F may not be reached	Unknown (known to operate at 32-125°F)	Unknown	Unknown	Unknown (known to operate at 14-104°F)

[a]Long-term testing has yet to be performed.

TABLE 3.6 Comparison of Various Detection Technologies to Address "Other Considerations" for the PETMAN Program

Consideration	Chemicapacitor	Fiber Optics	PADs	SAW
Immune to potential effects of perspiration or water vapor beneath garment ensemble to be tested	Yes	Likely[a]	Yes	Yes
Immune to potential interferents from off-gassing from polymers (from which garments may be made), decontamination chemicals, and cleansers	Yes	Likely[a]	Possibly no	Yes
Packaged in durable device that can withstand rough handling and repeated use	Yes	Yes	Yes	Yes
Portable enough to be conveniently placed in different location on test mannequin and beneath garment?	Yes	Yes	Yes	Yes[b]
Results in "false-positive" reading for test agents (GB, HD, or simulants)	No[c]	No[c]	Unknown	No

[a]Additional testing must be performed to confirm that sensor is not affected by perspiration or water vapor or by components that may off-gas from protective garment.
[b]Additional development work will be necessary to reduce "footprint" and profile of device.
[c]Additional testing must be performed to confirm that no "false-positive" readings occur.

nanotube,[3] and are capable of detecting organic vapors, [4] and inorganic gases—including CWAs and TICS.[5] Early sensors measured capacitive humidity.[6] Current commercially available chemicapacitor sensor arrays contain 8 to 16 sensors on a 2 × 5-mm chip, which allows for a small, portable detector package (with electronics measuring about 2.5 × 2 × 1 in.).[7] Key characteristics of chemicapacitor sensing based on the references cited are given below.

Dynamic Range and Detection Limit: Sensitivity to most VOCs is in the low parts per million (ppm) range. Limits vary with the vapor pressure of the target analyte.

Response Time: The mass-transfer limitations of the thick chemoselective dielectric material (~1 µm) may cause a sensor to appear slower than it is. Under optimal conditions in the laboratory, the chemicapacitors demonstrate a response time of less than 1 s and a recovery time of 1-2 s after exposure to many vapors.

Response over Temperatures and Humidities: Chemicapacitor sensors have been tested over a temperature range of 32-125°F and over the full range of relative humidity (0-100% noncondensing). As in all polymer-based technologies, sensitivity is a function of the temperature and vapor pressure of the target.

Fiber-Optics-Based Detection Technology

Fiber-optics-based chemical agent detection uses optical spectroscopy to detect chemicals absorbed to an active polymer-, porphyrin-, or enzyme-treated substrate.[8] Optical signals pass to a detector by way of

---

[3]Snow, E. S., F. K. Perkins, E. J. Houser, S. C. Badescu, and T. L. Reinecke. 2005. Chemical Detection with a Single-Walled Carbon Nanotube Capacitor. *Science* 307(5717):1942-1945.

[4]Patel, S. V., T. E. Mlsna, B. Fruhberger, E. Klaassen, A. Cemalovic, and D. R. Baselt. 2003. Chemicapacitive Microsensors for Volatile Organic Compound Detection. *Sensors and Actuators, B: Chemical* 96(3):541-553.

[5]Mlsna, T. E., S. Cemalovic, M. Warburton, S. T. Hobson, D. A. Mlsna, and S. V. Patel. 2006. Chemicapacitive Microsensors for Chemical Warfare Agent and Toxic Industrial Chemical Detection. *Sensors and Actuators, B: Chemical* 116(1-2):192-201.

[6]Delapierre, G., H. Grange, B. Chambaz, and L. Destannes. 1983. Polymer-Based Capacitive Humidity Sensor: Characteristics and Experimental Results. *Sensors and Actuators* 4:97-104.

[7]Seacoast Science, Inc.: *http://www.seacoastscience.com/Downloads/Seacoast_White_Paper_DEC%202006.pdf*, accessed August 28, 2007.

[8]Wolfbeis, O. S. 2004. Fiber-Optic Chemical Sensors and Biosensors. *Analytical Chemistry*. 76(12):3269-3283.

a single or bundle of optic fibers. The response of the detection system to the nonabsorbed radiation determines the types and amounts of chemical agents present. A key flexibility in the use of optical systems is the ability to incorporate optical fibers into the sensor system and allow one or more central sensor systems with fibers running to multiple locations in the mannequin.

These systems are used in variety of ways. The committee heard talks about two different fiber-optics-based chemical agent sensing approaches (see Appendix D). One way is with multiple chemical coatings (sensors) on bundled fibers that have differential partitioning of the target analyte.[9] Another way uses a single fiber interacting with a solid-state substrate.[10] Bundled fiber sensors can be constructed with thousands of fibers allowing the use of hundreds of coating replicates, which improve signal-to-noise ratios. In addition, a single coating can be placed over all the fibers in a bundle to improve sensitivity and reaction time. Key characteristics of fiber-optic sensing based on the references cited are given below.

Dynamic Range and Detection Limit: Some fiber-optics-based technologies can measure from 25 ng to over 1 μg of gas, but this can vary with the fiber-optics system.

Response Time: Under optimal conditions in the laboratory, the fiber-optics-based system has a response time of less than 1 s.

False-Positive Readings: Preliminary laboratory testing suggests the absence of false-positive readings. Additional laboratory testing is needed.

Response over Temperatures and Humidities: Preliminary laboratory testing suggests that fiber-optics-based systems can operate over a range of temperatures and humidities. Additional laboratory testing is needed.

---

[9]Albert, K. J., and D. R. Walt. 2003. Information Coding in Artificial Olfaction Multisensor Arrays. *Analytical Chemistry* 75(16):4161-4167; Epstein, J. R., M. Lee, and D. R. Walt. 2002. High-Density Fiber-Optic Genosensor Microsphere Array Capable of Zeptomole Detection Limits. *Analytical Chemistry* 74(8):1836-1840; Walt, D. R. 2000. Molecular Biology: Bead-based Fiber-Optic Arrays. *Science* 287(5452):451-452.

[10]White, B. J., and H. J. Harmon. 2005. Enzyme-Based Detection of Sarin (GB) Using Planar Waveguide Absorbance Spectroscopy. *Sensor Letters* 3(1-4):36-41; White, B. J., J. A. Legako, and H. J. Harmon. 2003. Extended Lifetime of Reagentless Detector for Multiple Inhibitors of Acetylcholinesterase. *Biosensors and Bioelectronics* 18(5-6):729-734; Legako, J. A., B. J. White, and H. J. Harmon. 2003. Detection of Cyanide Using Immobilized Porphyrin and Myoglobin Surfaces. *Sensors and Actuators, B: Chemical* 91(1-3):128-132; White, B. J., J. A. Legako, and H .J. Harmon. 2003. Spectrophotometric Detection of Cholinesterase Inhibitors with an Integrated Acetyl-/Butyrylcholinesterase Surface *Sensors and Actuators, B: Chemical* 89(1-2):107-111.

Surface Acoustic Wave (SAW) Sensors

SAW sensors are a mature technology first described in the 1970s.[11] Many advances in this technology have been made since then.[12] SAW chemical sensors are based on the change in the propagation of an acoustic wave over a piezoelectric substrate treated with a polymer coating. The polymer coatings or films are selected so that each will have a different chemical affinity for CWAs and TICs. In operation, SAW sensors are typically used in conjunction with a sample preconcentrator. Key characteristics of SAW technology based on the cited references are presented below.

Dynamic Range and Detection Limit: The dynamic range of a typical SAW chemical sensor is 5-6 orders of magnitude, measuring from about 1 pg to 1 µg of vapor. However, it depends on the vapor being measured. Nerve agents, such as GB, can usually be detected at less than 1 mg/m^3, whereas the detection limit for HD is about 1-2 mg/m^3.

Response Time: The time for the sensor to respond to mass changes in the selective polymer coating is usually less than 1 ms. In typical vapor-sensing applications, however, it is more likely that the response time will be determined by the time needed for the preconcentrator to obtain the sample, then for the vapor to be transported to the polymer-coating surface and for equilibrium to be established. The true response time could be greater than 10 s (with currently available systems).

False-Positive Readings: SAW devices are effective and reliable for detection of very small amounts of CWAs, such as GB and HD. SAW devices are typically not subject to false alarms due to the presence of other compounds in the sample stream. A preliminary separations stage and careful experimental procedure design should minimize the risk of false positives due to garment off-gassing or residual decontamination chemicals. If designed properly, a SAW detector can be used to detect CWAs effectively in a variety of environmental conditions.

Response Over Temperatures and Humidities: Although the polymer coatings used in SAW sensors can change physically when the devices are exposed to conditions outside the operating-temperature range, testing has

---

[11] H. Wohltjen, and R. Dessy. 1979. Surface Acoustic Wave Probe for Chemical Analysis I. Introduction and Instrument Design. *Analytical Chemistry* 51(9):1458-1475.

[12] See the following reviews for more information: Grate, J. W. 2000. Acoustic Wave Microsensor Arrays for Vapor Sensing. *Chemical Reviews* 10(7):2627-2648; Smith, J. P., and V. Hinson-Smith. 2006. Product Review: The New Era of SAW Devices. *Analytical Chemistry*, June 1, pp. 3505-3507.

confirmed that SAW sensors can maintain their accuracy in temperatures of 14-104°F. The various polymer coatings used in the SAW devices have various sensitivities to humidity, so a concentrator can be incorporated into the design to permit moisture to bypass the detectors while vapors of interest are analyzed.

DESIGN CHALLENGES AND HOW THEY MAY BE ADDRESSED

Although the technologies cited above have the ability to detect CWAs, meeting most of or all the requirements cited in Tables 3.5 through 3.7, several design challenges are related to their adaptation or incorporation into the PETMAN system. Those challenges are categorized as associated with optimization or tailoring of the basic sensing technologies for this application or with incorporation into a mobile PETMAN mannequin. The challenges are discussed below.

Challenges Associated with Validation and Optimization for the PETMAN System

The sensor systems that need to be validated and optimized for use in the PETMAN system have several aspects, such as size, time response, "recovery time," and hysteresis.

Size ("Footprint" and Profile): Obviously, some sensor technologies have a smaller footprint and profile than others. The "footprint" and thickness or volume should be minimized so that they do not interfere with the operation of the PETMAN or the "natural" movement of the garment. The PAD sampler is 1.5 × 1.25 × 1/8 inches. Current SAW devices are 3.5 × 2 × 1.2 inches and chemicapacitor devices are about 2.5 × 1.5 × 0.5 inches. However, the sensing component in these devices is only a fraction of the size, so it may be possible to separate the location where the analyte-coating interaction occurs (the sensor) from the electronics that power the unit to minimize the size of the devices. Fiber-optic sensors are likely to be much smaller and may be easier to incorporate into the PETMAN system.

Effects of Humidity: Testing is needed to ensure that the detector is not affected by condensed moisture (for example, associated with perspiration).

Time of Response: It is estimated that SAW sensors can make readings every 10 s or more. That meets the threshold requirement (T), but additional SAW sensor development may be required if the stated objective (O) of a reading every 1 s is to be achieved. The chemicapacitor and fiber-optics-based sensors are expected to have response times of 1 s or less. If a preconcentrator or an initial separation stage is required, this time required for analysis could increase.

Effects of Temperature: The SAW sensor will provide an accurate estimate of the amount of agent present over a temperature range of 14-104°F. Testing will have to be performed to determine the influence of temperature on the performance of the fiber-optics-based and chemicapacitor sensors. If the stated PETMAN objective requirement of 125°F is to be achieved, a major R&D effort for sensor-coatings could be necessary.

Reversible Response of Sensors: Developers of chemicapacitor and SAW detection technologies have stated that their technologies are reversible, that is, the sensing units will detect a drop in the concentration of CWAs. But some fiber-optics-based detection systems may not be able to detect a drop in the concentration of agent (after exposure to a higher concentration), and further development may be necessary to overcome that deficiency.

Hysteresis: Additional testing will be necessary to assess how many times the coatings of the various sensors can be used. Alternatively, it may be more feasible simply to replace "coated component" parts, such as those used with the fiber-optics-based systems, after each test.

Challenges Associated with Incorporation into the PETMAN

A number of design considerations are related to physical incorporation of detection technology into the PETMAN system, for example, mounting, location, robustness, and power source.

Mounting of Sensor on PETMAN: It will be a challenge to provide an unobtrusive package that can be mounted at multiple points (50 or more locations) on the mannequin body. The fiber-optics-based sensors appear to be small enough to be mounted in a nearly flush package (less than 0.1 inch high) or possibly woven in a manner similar to the Georgia Institute of Technology adaptive and responsive textile structure (ARTS) fabric system. (See Chapter 5 for a description of this technology.) If SAW or chemicapacitor technology is used, however, the mannequin may require recessed areas to accommodate and minimize the profile of the individual sensing devices.

Resistance to Shock and Vibration: To ensure that a sensor's response is not affected by exposure to shock and vibration, it may be necessary to envelop it in an inert damping polymeric material, such as silicone sponge.

If so, damping should not interfere with agent flow to the sensor or with the sensor's response to the agent.

Powering (Wireless vs Hard Wiring or Tethering of Sensing Units): Multiple options are available for powering the sensor technology. For example, if internal wiring can be made available, a socket capable of receiving a standard dual-in-line package (DIP) could be built into the skin of the mannequin and even recessed to allow a flush surface. The wires under the skin could connect to a readout circuit elsewhere on the body and thereby connect to a wireless interface.The sensor chip could be tethered to the readout circuit via a flexible printed-circuit cable (also called laminated flat flexible cables). The sensor package would be less than 0.1 inch high, and the circuit board could be in a part of the mannequin where it would not interfere with movement. Flexible wiring (less than 0.04 in. thick) would connect to a readout circuit elsewhere on the body and thereby connect to a wireless interface.

Protection from Condensed Moisture or Perspiration: It may be necessary to protect (by design or positioning) the sensors from direct exposure to or contact with condensed water associated with perspiration. Again, interference with agent flow or sensor response must be ensured.

Airflow Across Sensing Units: Sampling analysis may require that air containing an analyte be pulled into and across a sensor's interaction surface. For example, a fiber-optics-based system may passively absorb and detect the presence of an agent, whereas a SAW or chemicapacitor sensor may require a sample to be pulled into its detection system. If the latter is the required, provisions must be made for creating airflow.

Decontamination or Replacement of Sensors: If it is necessary to decontaminate or replace a sensor or sensor component after each test, provisions must be made for easy access.

Signal Interference: If the system is wireless, it must be ensured that transmitted signals from other components of the PETMAN do not interfere with sensor signals.

COST-BENEFIT ANALYSIS AND TRADE-OFFS

The chemicapacitor, fiber-optics-based, and SAW technologies have the potential to meet the cited requirements for the detection of CWA penetration in the PETMAN system. It is likely that all the sensor technologies will require additional validation, optimization, and modification.

The hurdles faced by organizations in developing and commercializing a new sensor can be formidable (Table 3.7). For example, although SAW and chemicapacitor technologies have matured to the point where commercially available detection systems are available, their current footprints and profiles may have to be reduced to be practical for PETMAN application. Although preliminary testing of fiber-optics-based technology suggests that it has great potential for this application, additional laboratory validation and modification for successful incorporation into the PETMAN system will be necessary.

It is difficult to accurately estimate the costs associated with the development and optimization of technologies for the PETMAN application. However, the cost to equip a PETMAN with a fiber-optics-based system (a finished product, excluding development and optimization costs) should be relative low, possibly less than $5,000.[13] If a SAW or chemicapacitor system is used, the product cost after development could exceed $10,000.[14]

FEASIBILITY AND POTENTIAL ALTERNATIVES

When assessing the feasibility of a given technology, the following were considered:

- Availability and reliability of the technology
- Ability to meet the PETMAN requirements, outlined in Tables 3.5 through 3.7
- Likelihood of overcoming the design challenges associated with the adaptation of the technology and its incorporation into the PETMAN system
- Time to develop and optimize
- Cost
- Maintenance

The three example technologies discussed—chemicapacitor, fiber-optics-based, and SAW—have the potential to meet the cited requirements for the detection of CWA penetration or permeation in the PETMAN. However, further study, optimization, and design challenges must be addressed before any of those technologies can be effectively used in the PETMAN mannequin application.

Alternatively, the PADs currently used in the MIST system for the cumulative detection of methyl salicylate may be adapted for this appli-

---

[13] Based on presentations by Harmon and Walt in Appendix D.

[14] Based on currently available chemical sensor systems from Mine Safety Appliances, Seacoast Science, and others.

TABLE 3.7 Commercialization of a New Sensor

Phase	Estimated Time to Complete (months)	Estimated Investment ($ millions)	Milestone
1	12-36	—	Laboratory prototype
2	4-6	0.3-0.5	Development and commercialization plan
3	6-10	0.5 -1	Final design
4	6-10	0.5-1	Manufacturing prototype
5	4-6	0.5-1	First production run
6	2-4	0.2-0.5	Market release
Total	22-36	2-4	

SOURCE: Taylor, R. F., and J. S. Schultz, eds. 1996. *Handbook of Chemical and Biological Sensors*, Institute of Physics Publishing, Philadelphia, Table 23.9, p. 573.

cation. Although it is low in cost and could be adapted for the detection of CWAs, the desired advances associated with the PETMAN application—such as detection of penetration or permeation of the garment in 1-s increments—would not be realized. On the basis of consideration of the material presented, the following conclusions were made:

Conclusion 3-1: Both the PADs currently used and the real-time sensors identified have the potential to achieve the PETMAN threshold requirements. The real-time sensors identified also have the potential to achieve the objective requirements of PETMAN. However, several design challenges exist for adapting or incorporating any of these sampling systems into the PETMAN. These challenges include:

- Mounting of the sensor onto the mannequin
- Resistance to shock and vibration
- Powering of the sensor units (wireless versus hard wiring or tethering of sensing units):
- Protection from condensed moisture (perspiration)
- Airflow across the sensing units
- Decontamination and/or replacement of the sensors
- Signal Interference

Recommendation 3-1: The existing set of PETMAN performance requirements and the additional criteria identified and described in this chapter should be used for the selection and evaluation of sensors for the PETMAN system.

Conclusion 3-2: If the PETMAN must operate in the environmental chamber at 125°F (an objective level requirement), a major R&D effort for sensor-coatings could be necessary.

Conclusion 3-3: The hurdles faced by organizations in developing and commercializing a new sensor can be formidable. The cost associated with the development and optimization of technologies for the PETMAN application could be millions of dollars, and the product cost after development could range anywhere from a few thousand dollars to more than $10,000.

4

Design Challenge: Robotic Capability for PETMAN

This chapter describes the specific robotic capability needed for the PETMAN system, including motion and articulation, power, and heating and cooling, according to the following PETMAN design challenges:

3.2.1 The study will determine the feasibility of designing a PETMAN system to be tethered (T) or free standing/self-contained (O). A tethered PETMAN system design must not compromise the integrity of the individual protection ensemble equipment being tested on the PETMAN system. If a tethered design is selected the design must also minimize the impact to the whole ensemble operation.

3.2.2 The study will determine the feasibility of designing a PETMAN system to be compatible with all individual protection and ancillary equipment as well as weapon systems defined in 3.3.9-3.3.10.4. Areas to be addressed are donning/doffing and proper size/fit of the individual protection equipment. The PETMAN system design shall meet the appropriate 50th percentile male anthropometric measurements, as defined in DOD-HDBK-743A, Military Handbook Anthropometry of U.S. Military Personnel, to allow for the necessary fit/seal that each piece of protective equipment requires.

3.2.5 The study will determine the feasibility of designing a PETMAN system that can simulate fixed skin temperature (by body region), perspiration rate (by body region), and respiration rate (T) and a realistic variability in skin temperature, perspiration rates, and respiration rates based on the amount of physical activity/exertion (O) defined in 3.3.4.1-3.3.4.3.

3.2.7 The study will determine the feasibility of designing a PETMAN system that can perform the Man-in-Simulant Test (MIST) exercises defined in 3.3.8 versus all motions. The study will determine the feasibility of the PETMAN system performing the human-like movements utilizing the degrees of freedom (DoF) and considerations defined in 3.3.7.

3.2.8 The study will determine the feasibility of designing a PETMAN system with fully articulated hands and feet that simulate human motion, the minimum amount of hand and foot articulation required for the PETMAN system operation and a partial level of hand and foot articulation.

RELEVANT PETMAN REQUIREMENTS

Additional relevant PETMAN requirements are presented in Box 4.1. The next section will describe current technology and the feasibility of meeting the PETMAN system requirements.

CURRENT TECHNOLOGY

Humanoid robots are those whose body structure resembles that of a human. The first humanoid robot was perhaps the WABOT-1 (Figure 4.1), developed by Kato's group at Waseda University in 1970-1973.

Several humanoid robots have since been developed, usually with a specific purpose in mind, such as to study bipedal locomotion, for entertainment, and to assist humans. In 2000, research and development in this field reached a higher level of sophistication when Honda introduced ASIMO.[1] The current version of ASIMO, released in December 2005, is a 34-degrees-of-freedom (34-DoF) robot. It can run, walk smoothly, climb stairs, communicate, and recognize people's voices. ASIMO is 130 cm tall, weighs 54 kg, and can operate for 1 hour on an internal battery.

A more recent robot is HRP-2,[2] which was also developed, by Kawada Industries, to assist humans. Videos of this model walking, wearing a suit, lying down and getting up, and performing such complex behaviors as dancing are readily available on the Web (for example, see *http://www.plyojump.com/hrp.html*). This robot is 154 cm tall, weighs 58 kg, has 30 DoFs, and can operate for about an hour on an internal NiMH battery. The current model, HRP-3P, has the same capabilities, six more DoFs, and

[1]ASIMO: The Honda Humanoid Robot. *http://world.honda.com/ASIMO/*. Accessed June 14, 2007.

[2]Humanoid Robot HRP-2 "Promet." Kawada Industries, Inc. *http://www.kawada.co.jp/global/ams/hrp_2.html*. Accessed June 14, 2007.

BOX 4.1
Relevant PETMAN Requirements

3.3.2 The PETMAN system shall be capable of operation for twelve (12) hours prior to requiring operational maintenance, three (3) months prior to preventive maintenance and six (6) months prior to calibration. (T) The PETMAN system shall be capable of operation for twenty-four (24) hours prior to requiring operational maintenance, six (6) months prior to preventive maintenance and twelve (12) months prior to calibration. (O) Operational maintenance is defined as the required maintenance procedures to prepare the PETMAN system for each test trial, for example, filling a perspiration reservoir, changing agent samplers or decontamination, before the next trial. Preventive maintenance is defined as a maintenance event performed prior to a failure in order to prevent its occurrence.

3.3.10 The PETMAN system shall be compatible with the following weapons systems. The PETMAN system must be able to hold/grip and aim the weapon IAW the field manual (FM) FM 3-22.9, *Rifle Marksmanship Field Manual.*

3.3.10.1	M4 Modular Weapon
3.3.10.2	M24 Sniper Rifle
3.3.10.3	M16A2 Rifle 5.56 MM
3.3.10.4	XM8 Lightweight Assault Rifle

3.3.13 The PETMAN system shall record the following system parameters over time: skin temperature, respiration rate, perspiration rate, and total mass (in nanograms) of chemical vapor that penetrates/permeates through the protective ensemble. The PETMAN system shall record the start and stop time of each motion in 1-second increments.

the ability to adjust to slippery surfaces when walking.[3] HRP-3 is 160 cm tall and weighs 65 kg.

Hubo, a robot developed at the Korea Advanced Institute of Science and Technology (KAIST), is similar in size (56 kg and 125 cm), shape, and function to ASIMO. It is a 41-DoF robot and can operate for about 1 hour on a Li-polymer battery.[4]

All the above examples use electric motors for actuation. Humanoid robots developed by Sarcos use hydraulic actuation. Some of the differences between the two methods are described later in this chapter. Robots such as

[3]This site is in Japanese: HRP-3P. Kawada Industries, Inc. *http://www.kawada.co.jp/mechs/hrp-3p/index.html*, *http://pc.watch.impress.co.jp/docs/2005/0909/hrp3.htm.*

[4]Humanoid Robot Research Center. Hubo Lab. http://www.hubolab.com/. Accessed June 14, 2007.

FIGURE 4.1 WABOT-1, developed at Waseda University in the early 1970s, was the first humanoid robot.
SOURCE: WABOT-1, Humanoid Robotics Institute, Waseda University.

the Sarcoman and the Sarcos Animation Figure are mounted so that they do not have to balance.[5] Numerous videos on the Internet demonstrate the ability of Sarcoman to perform a variety of tasks that require dexterity. These types of robots have been used to test spacesuit designs, in the entertainment industry, and in robotics development and research laboratories.

Dexter, a system designed by Anybots,[6] and Lucy, a system still under development at Vrije Universiteit Brussel,[7] both operate with pneumatic actuation. This type of actuation mechanism is still being developed. Dexter is 175 cm tall, weighs 61 kg, and does not have arms. Lucy is only a pair of 150-cm-tall, 30-kg legs. The goal of these robots is to study the use of pneumatic actuation in bipedal walking robots.

Robonaut, a teleoperated robot developed by NASA,[8] was designed to work with astronauts in zero-gravity environments. Inasmuch as astronauts' legs generally work as one unit in space, developing a system that could walk was not necessary. Instead, dexterity in the upper body was considered of paramount importance. The most recent version of Robonaut, Robonaut R1b, is essentially an upper body that can be mounted either on one "leg", on a two-wheel base, or on an all-terrain vehicle appropriate for extreme environments. Robonaut is intended to work cooperatively with humans, be controlled by humans, and use tools designed for humans, so its human-size hands are complex, with five fingers, 12 DoFs in the hand itself, and two DoFs in the wrist. The development of the hand was closely tied with the development of the arm and would not be easily transferable to a different system.

The state-of-the-art humanoid technology is impressive. It can readily perform many simulated exercises. However, none of the humanoid robots reviewed above is suited for PETMAN applications, although they constitute a technologic basis for what is or is not feasible. The major limitations of the current humanoid robots for PETMAN applications are of four types:

- *Proportion.* Most humanoid robots look like a man in a space suit. They have been adjusted to human proportions in the vertical direction (such as height, foot length, and arm length), but their legs and arms are too fat to accommodate joint actuation mechanisms.

[5] Tele-robotics: High Performance Humanoid Robot. Sarcos. *http://www.sarcos.com/telespec.atr.html*. Accessed June 14, 2007.

[6] "About the Robots." Anybots. *http://www.anybots.com*. Accessed June 14, 2007.

[7] Bipedal Robot Walking Lucy. Department of Mechanical Engineering, Vrije Universiteit Brussel. *http://lucy.vub.ac.be/index.htm*. Accessed June 14, 2007.

[8] December Tests at JSC. Robonaut. NASA Johnson Space Center. *http://robonaut.jsc.nasa.gov/*. Accessed June 14, 2007.

- *Operating time.* Electric humanoids look attractive for PETMAN applications because they are battery-operated and obviate a power cable, but the continuous operating time is limited to about an hour.
- *Robustness.* Current humanoid robots are not designed for robustness. Data are not available to estimate the reliability of current technology for PETMAN applications.
- *Primitive hands.* Most humanoid robots have too few degrees of freedom for hands. Attempts to introduce additional degrees of freedom will make the arms heavier and possibly fatter.

Movement and Degrees of Freedom

Current humanoid robots typically have lower maximal velocities than humans. For example, most of the joints of H7, a bipedal robot developed at the University of Tokyo, are limited to about 1 revolution/s.[9] In addition, the range of motion of most joints is less than a human's. Tables 4.1 and 4.2 give ranges of motion of HRP-2 and Hubo. Note that the human ankle has a yaw degree of freedom not listed in these tables.

In light of the available technology, the PETMAN requirements describing degrees of freedom are too vague and do not highlight the specific needs of the project. We recommend that the sponsor add specific range-of-motion and maximal-velocity requirements for the desired behaviors to any broad agency announcement (BAA). Motion-capture studies of movements and donning and doffing of soldiers performing MIST exercises should be used to refine BAA requirements for range of motion and maximal joint velocities and to estimate forces and torques necessary for the desired mannequin behaviors. Motion-capture studies are often carried out in this way to understand complex movement such as in medical or sports science.[10]

Actuation and Transmission Technology

The technology used for actuation of the PETMAN system will be a key feature of any development plan. The type of actuation in a robot affects range of motion, sizes of limbs and joints, power use, heat generation and

[9]See JSK website: *http://www.jsk.t.u-tokyo.ac.jp/research/h6/H6_H7.html.* Accessed October 9, 2007.

[10]Brady, R. 2000. Foot Displacement but Not Velocity Predicts the Outcome of a Slip Induced in Young Subjects While Walking. *Journal of Biomechanics* 33(7):803-808; Brown, W. M., L. Cronk, K. Grochow, A. Jacobson, C. K. Liu, Z. Popovič, and R. Trivers. 2005. Dance Reveals Symmetry Especially in Young Men. *Nature* 438:1148-1150; Davis, I. 2006. Biomechanical Factors Associated with Tibial Stress Fracture in Female Runners. *Medicine and Science in Sports and Exercise* 38(2):323-328.

TABLE 4.1 Ranges of Motion of the HRP-2 Robot

Joint			(a) Standard Human	(b) HRP-2
Head		R	-50 deg. to 50 deg.	no existence.
		P	-50 deg. to 60 deg.	-30 deg. to 45 deg.
		Y	-70 deg. to 70 deg.	-45 deg. to 45 deg.
Right Arm	Shoulder	R	-90 deg. to 0 deg.	-95 deg. to 10 deg.
		P	-180 deg. to 50 deg.	-180 deg. to 60 deg.
		Y	-90 deg. to 90 deg.	-90 deg. to 90 deg.
	Elbow	P	-145 deg. to 0 deg.	-135 deg. to 0 deg.
		Y	-90 deg. to 90 deg.	-90 deg. to 90 deg.
	Wrist	R	-55 deg. to 25 deg.	no existence.
		P	-70 deg. to 90 deg.	-90 deg. to 90 deg.
Right Hand		P	0 deg. to 90 deg.	-16 deg. to 60 deg.
Waist		R	-50 deg. to 50 deg.	no existence.
		P	-30 deg. to 45 deg.	-5 deg. to 60 deg.
		Y	-40 deg. to 40 deg.	-45 deg. to 45 deg.
Right Leg	Hip	R	-45 deg. to 20 deg.	-35 deg. to 20 deg.
		P	-125 deg. to 15 deg.	-125 deg. to 42 deg.
		Y	-45 deg. to 45 deg.	-45 deg. to 30 deg.
	Knee	P	0 deg. to 130 deg.	0 deg. to 150 deg.
	Ankle	R	-20 deg. to 30 deg.	-20 deg. to 35 deg.
		P	-20 deg. to 45 deg.	-75 deg. to 42 deg.

NOTE: R: Roll axis, P: Pitch axis, Y: Yaw axis

SOURCE: Kaneko, K., F. Kanehiro, S. Kajita, H. Hirukawa,T. Kawasaki, M. Hirata, K. Akachi, and T. Isozumi. Humanoid Robot HRP-2. Proceedings of the 2004 *IEEE International Conference on Robotics & Automation*, New Orleans, LA. April 2004. © 2004 IEEE.

TABLE 4.2 Ranges of Motion of the Hubo Robot

		Angle Range
Hip	Yaw	0 ~ +45°
	Roll	–31° ~ +28°
	Pitch	–90° ~ +90°
Knee	Pitch	–10° ~ +150°
Ankle	Pitch	–90° ~ +90°
	Roll	–23° ~ +23°

SOURCE: Presentation by Jun-Ho Oh, April 2, 2007.

distribution, deliverable power and control, ease of joint movement when the system has no power (backdriveability), and more. In practical terms for the PETMAN system, the method of actuation will affect the ability of the robot to perform exercises with the desired forces and range of motion, the ease of maintaining a stable temperature, the likelihood of achieving anthropomorphic dimensions, the ease of donning and doffing the individual protection ensemble (IPE), and the need for a tether.

Several actuation technologies, including electric, hydraulic, and pneumatic systems were considered, but there is no clear leading technology. Most commercial humanoid robots use electric torque motors and harmonic drives, but there are many other possibilities, including:

- Electric systems with battery power
- Electric systems with a tether to deliver power
- Hydraulic systems with an onboard battery-powered hydraulic pump
- Hydraulic systems with an onboard hydraulic pump and an external power source (electric tether)
- Hydraulic systems with an external hydraulic pump (pressurized hydraulic fluid in the tether)
- Pneumatic systems with an onboard battery-powered air compressor
- Pneumatic systems with an onboard air compressor and an external power source (electric tether)
- Pneumatic systems with an external air compressor (pressurized air in the tether)

Advantages and disadvantages of various systems are described in Table 4.3.

All-electric systems exist, but the overall perspective that emerged in our discussion is that hydraulic and pneumatic systems may also be feasible.

TABLE 4.3 Advantages and Disadvantages of Electric, Hydraulic, and Pneumatic Actuation

Type of Actuation	Advantages	Disadvantages
Electric	- Precisely controlled, high- control bandwidth - Can be sealed and cleaned. - Rotary actuation (compact design is possible for some joints) - Motor-reduction gear set-harmonic drive can produce large torques - Joint torque feedback can be used to achieve backdriveability and good torque control	- Inappropriate form factor for some joints (ankle, wrist, fingers). - Complicated design for linear actuation - Requires heat dissipation - Extensive gearing is typical with poor backdriveability and little torque control
Hydraulic	- Can produce large forces - Linear actuation can be appropriate for some joints - Precisely controlled high- control bandwidth -Joint torque feedback can be used to achieve backdriveability and good torque control	- Requires hydraulic fluid lines, with space requirements - Needs to handle hydraulic-fluid leakage - Requirse hydraulic pump - Inappropriate form factor for some joints; complicated design for rotary actuation - Pump may need heat dissipation - Difficult to achieve human range of motion with rigid linear actuators
Pneumatic	- Compact - Lightweight - Inherently backdriveable	- Pneumatic system must be closed; air cannot be exhausted into ensemble - Difficult to control because of air compressibility and transmission delay; low-control bandwidth - Requires air lines and air compressor with associated space requirements - Range of motion may be low - Bulky size or high air pressure needed for large force - Inappropriate form factor for some joints; complicated design for rotary actuation

To avoid the use of a tether, an onboard hydraulic pump or air compressor and associated power source may need to be developed.

Ability to Don and Doff without Modification of PPE

Although the robot does not need to don and doff the protective gear autonomously, it does need to be possible for a human to put the protective gear on the robot and take it off. That may require that the robot be backdriveable or that it be able to move in such a way as to assist in donning and doffing.

Power and Energy

The threshold power requirements for the PETMAN system call for a tethered system. With a power line attached to the robot providing unlimited, uninterrupted power, it would be feasible to meet either the 12-h (T) or 24-h (O) operational maintenance requirement. Without a tether (O), however, feasibility is more difficult to assess. Examples of other robotic systems that perform legged locomotion suggest power requirements of around 300 W at the battery terminals during moderate walking.[11] ASIMO uses a NiMH 38.4-V 10-Ah battery; HRP-2 uses a NiMH 48-V 14.8-Ah battery; and Hubo uses a 24-V, 20-AhLi-polymer battery.[12] Use of Li-polymer batteries at 150 W-h per kilogram would require batteries weighing 2 kg for an hour of operation. New models of ASIMO run for 1 h between charges.[13] Hubo runs for about 60 min between charges.[14] Considering only the power requirements for motion and actuation (not including simulation of human physiology, sensors, and so on) and a test schedule that has the robot just standing at least 50 percent of the time, a two hour untethered test seems feasible with current technology before charging is required.

Although plugging the mannequin into the wall (or plugging a cable into the mannequin) is the most obvious method of recharging the system, other possibilities could be considered:

---

[11]Based on Dr. Saoshi Kagami's humanoid, which consumes 300-400 watts (100 watts to power the CPU), weighs 58 kg, is 1.58 m tall, and walks at 2.1 km/hr. See Bekey, G., Ambrose, R., V. Kumar, A. Sanderson, B. Wilcox, and Y. Zheng. 2006. WTEC Panel Report on International Assessment of Research and Develoopment in Robotics, World Technology Evaluation Center, Baltimore, Md., page 87.

[12]Jun Oh's talk to committee.

[13]Honda ASIMO. *http://asimo.honda.com/InsideASIMO.aspx*. Accessed October 9, 2007.

[14]Humanoid Robot Research Center. Hubo Labs. *http://www.hubolab.com*. Accessed June 14, 2007.

- It is feasible to recharge a battery through electric contacts on the boot or on the face plate to allow longer tests.
- It is feasible to charge the robot inductively to allow longer tests.

Any method of recharging should be evaluated in light of any necessary IPE modifications and the length of time needed to recharge the system fully.

Heating and Cooling

Current humanoid robots may produce 100-200 W of heat more than a human being. In the IPE, removing that heat will be difficult. For the mannequin to maintain "skin" temperatures similar to those of a human, the system will most likely require heating and cooling at different points in the operation. In addition, motors, actuators, and other components may create hotspots in the robot, and the extremities will probably be cooler. An electric system seems to be susceptible to severe overheating in an enclosed ensemble, and some active means for circulating coolant through internal instrumentation and to actuators in the limbs may be needed. For example, Hubo uses fans for cooling the main computer, control boards, and DC motors, and HRP-2 has a cooling system for the leg actuators. A hydraulic system may not require cooling but would still need temperature regulation to simulate skin temperature. An onboard hydraulic pump and its power supply will probably need to be cooled. Thus, to maintain a human-like skin-temperature gradient, it will probably be necessary to develop both heat-distribution and heat-dissipation or cooling systems.

Having considered various cooling systems, it is feasible to develop distribution and cooling systems for integration into a test mannequin.[15] For example, it may be possible to preload the system with a disposable coolant, such as liquid nitrogen or solid CO_2, with a circulating liquid and vent exhaust gases as part of the breathing simulation. Spacesuit technologies may be adaptable to the mannequin's needs, but they usually require the presence of a convenient radiator. A tethered cooling system may be useful for all actuation types.

Tethered or Untethered

As discussed above, the combination of power and actuation demands and the volume constraints of the human body shape suggest that a tethered system may be a useful option for using current technologies. Once a tether is introduced, it might be used for electric power, hydraulics, communica-

[15]See Chapter 2.

tion and sensing signals, and heating and cooling. However, the integration of the tethered system has two major implications:

- *Penetration of the ensemble.* The tether must pass through the protective ensemble at some point and would therefore typically require custom modification of each ensemble used in each test. In addition, the presence of the tether may introduce some abnormal behavior of the ensemble fabrics and materials that could alter the pattern of penetration.
- *Forces induced by the tether.* The tether must be designed to minimize mechanical forces exerted on the mannequin due to the connection of the tether. Some existing technologies are able to support the weight of a tether and also may be used to move a tether, reducing unnatural dynamic forces that may result.

The presence of a tether raises the issue of where to place the connection. From a broad perspective, the head (helmet) and the back might be logical sites, but both may require modification of important regions of the ensemble and may influence critical performance of the ensemble and air motion. An alternative is to consider the foot (heel or sole). The boot would need to be modified, but perhaps a smaller range of boot designs could be an acceptable trade-off. A tether on the back of the heel might minimize interaction with the mannequin motion and still allow normal foot placement in standing, walking, jumping, and crawling tests.

In summary, we believe that a tether incorporating electric power, coolant, and command signals could be flexible and run through the face plate or the side of the boot to allow longer tests. It may be possible to design a mannequin for which a tether is used only on longer tests.

Motion and Control

Simulation and monitoring of the Man-in-Simulant Test (MIST) exercises by the mannequin will require movement programming and movement data recording, both of which are well-established capabilities. For example, programming of humanoid dancing by Ikeuchi's group is an example of behavior control similar to what is needed for this project,[16] and walking, climbing a ladder, sitting, lying down, getting up, and use of small weights have all been demonstrated by robots mentioned earlier in this chapter. As mentioned earlier, many of the required motions could be programmed with the aid of motion-capture technology.

[16]Computer Vision Lab: IKEUCHI Laboratory, Institute of Industrial Science, University of Tokyo. *www.cvl.iis.u-tokyo.ac.jp*. Accessed June 14, 2007.

However, for the PETMAN system in particular, specific technical issues should be considered. The control system's need to handle different weights and weight distributions of different ensembles and carried weapons increases the complexity of the required motion programming. It may also be necessary to adapt to the load distribution of water or sweat as it moves from storage areas to the boots and gloves.

In addition, because of the human shape and function requirements of the system, there is concern about whether the torques required for jumping jacks, marching, and crawling are achievable with current actuators and transmissions. The recommended motion-capture study should highlight any limitations in this regard. Also, proposing groups should be encouraged to consider a variety of actuation and transmission types—including electric, hydraulic, and pneumatic, perhaps in combination—to accommodate the full range of required motions.

Anthropomorphic Form and Function

The PETMAN anthropomorphic shape requirements place severe constraints on the volume of internal systems—such as power, cooling, computing, and communication—and on the mechanisms themselves, including actuators and associated gearing. Review of current technologies suggests that both constraints will present challenges to the PETMAN designers.

Packaging power, cooling, computing, and communication in the torso will be difficult with today's technologies for an untethered system. As mentioned above, the batteries for self-contained operation for the anticipated maximal time of 1-2 h would have substantial weight and volume. Similarly, a computer system, communication electronics, sensor instrumentation, gas exchange, heating-cooling apparatus, and all attendant "plumbing" would need to be accommodated.

The current generation of self-contained humanoid robots with electric motors all exceed the proportional volume and shape requirements of a mannequin because of the bulk of electric actuators and gears. Specifically, it may be difficult to reduce the size of the ankle and wrist to human size and form with electric rotary actuators and harmonic drives (the current technology). These systems also have relatively simple hands and feet and therefore do not meet the functional requirements of PETMAN. The use of hydraulic systems could reduce the bulk of actuators and gearing and meet the form requirements. Similarly, some combination of electric and hydraulic actuators might be engineered to achieve a compromise.

Control poses another important challenge to achieving human dimensions. Current control approaches, such as zero moment point (ZMP) control, make use of a wide foot. It may be difficult to modify those approaches in the period allocated for development.

Hands and Feet

The PETMAN requirements for human-like hands and feet integrated into a humanoid robot are beyond the state of current technology. Designing a robotic hand that fully mimics a human hand is quite difficult, and current robot hands lack human degrees of freedom, range of motion, and speed.[17] Available models would be difficult to integrate into a new system, are larger than an average human male hand, and do not meet the threshold (T) requirements of PETMAN. Substantial additional research would be required to develop fully human-like hands as well as feet for a PETMAN system.

However, more limited hands and feet are available, and the current requirements go beyond the needs of the project. A useful reference point is the National Aeronautics and Space Administration (NASA) Robonaut system,[18] which uses articulated thumbs and fingers to perform basic grasping and manipulation functions but lacks a number of the specific motions of a human hand. We recommend that a motion-capture study be performed to determine the hand functions required for PETMAN. For feet, a simple study of foot mockups with toe joints should be performed to determine the joints and flexibility necessary to don and doff boots and to accomplish other desired movements.

Reliability and Maintenance

Evaluating the reliability and maintenance schedule of a PETMAN system is difficult for several reasons. It is not clear what the wear and tear characteristics of the PETMAN tests will be. Reliability has not been emphasized in humanoid-robot research, so the degree to which components will need to be oversized to make current robot designs reliable is not known; this may have implications for system size, weight, and thermal issues. Finally, given the small number of humanoid robots in existence, we do not have extensive data on maintenance requirements.

COST-BENEFIT ANALYSIS

Developing a PETMAN system that can perform all the desired motions will require substantial modifications of current humanoid-robot designs. Current commercially available humanoid robots cost about $1 million. The estimated cost of developing the motion system of PETMAN would be

[17]Shadow Dexterous Hand, Shadow Robot Company, *http://www.shadowrobot.com/hand/* Accessed October 9, 2007.

[18]Robert Ambrose, National Aeronautics and Space Administration, presentation to committee (see Appendix D).

about $10 million. The proposer should be required to address the serious risks associated with the use and development of this system:

- The robot falls down and requires expensive repair.
- The robot is expensive to decontaminate (see Chapter 5).
- The robot cannot be packaged in human form.
- Operation for 1-2 h may not be long enough for the sponsor's needs, but longer tetherless runs cannot be achieved.

FEASIBILITY AND POTENTIAL ALTERNATIVES

In light of the full list of requested functions, it will be difficult to design an untethered (objective requirement), free-standing, 50th percentile robotic test mannequin with an operational time of two hours. The main design challenge will probably be the integration of all systems (power, control, sensors, respiration, actuation, and so on) within the size constraints; this is discussed in some detail in Chapter 7. Meeting the threshold requirement of a tethered system, which would reduce the number of systems housed in the mannequin, is feasible. It may also be possible to design a system that allows two operational modes: one untethered that meets some of the performance requirements and operational time, and one tethered with greater performance options that is not constrained by battery life and the size of coolant and perspiration reservoirs. Performance might be improved if the sponsor adjusted the design requirements to accommodate only the prescribed motions of the test rather than fully human-like motions (limit DoFs, range of motion, maximal velocities, and so on).

With respect to motion and control of the PETMAN system, the conclusions and recommendations are as follows:

Conclusion 4-1: Achievement of the full human range of motion and speed will require the development of new actuators or transmissions. In light of the currently available technology, the PETMAN requirements regarding degrees of freedom are too vague and do not represent the specific needs of the project.

Recommendation 4-1a: Specific range-of-motion and maximal-velocity requirements for the desired behaviors should be included in any broad agency announcement (BAA).

Recommendation 4-1b: Motion-capture studies of test movements and of donning and doffing should be performed to refine BAA requirements regarding range of motion and maximal joint velocities. This will also be useful for estimating forces and torques necessary for the desired behaviors.

An additional motion-capture study should be performed to determine necessary hand function required for PETMAN. For the feet, a simple study of foot mockups with toe joints should be performed to determine the foot joints and flexibility necessary to don and doff boots and to accomplish other desired movements.

Conclusion 4-2: Achievement of tests longer than two hours will require either limited motion, a recharge-refueling method, a tether, or research on new power sources and actuators With a power line attached to the robot providing unlimited, uninterrupted power, it would be feasible to meet either the 12-h (threshold) or 24-h (objective) operational maintenance requirement. Without a tether, however, the feasibility is more difficult to assess.

Conclusion 4-3: It is feasible to develop current distribution and cooling systems for integration into a testing mannequin, but improvement of thermal-regulation technologies (heat distribution, cooling, and so on) will probably be required, especially for an untethered design. The problem is simpler if a tether is available to provide a convenient line for heating and cooling purposes.

Conclusion 4-4: Achievement of the human form incorporating all desired features may not be possible with current actuation technology. In particular, it may be difficult to reduce the ankle and wrist to human size and form with electric rotary actuators and harmonic drives (the current technology). Modification of current motion-control approaches may be necessary. The feasibility would increase if a way to run a hydraulic compressor electrically, either with a battery or with electricity delivered through a tether, were developed or if high-performance pneumatic control were demonstrated.

Conclusion 4-5: Given the specific, unique requirements of PETMAN and the small number of humanoid robots in existence, there are not enough data available to assess the reliability of the system.

5

Design Challenge: PETMAN Surface Structure and Materials

This chapter addresses the feasibility of designing the mannequin surface or "skin" according to the following PETMAN design challenges:

3.2.2 The study will determine the feasibility of designing a PETMAN system to be compatible with all individual protection and ancillary equipment as well as weapon systems defined in 3.3.9-3.3.10.4. Areas to be addressed are donning/doffing and proper size/fit of the individual protection equipment. The PETMAN system design shall meet the appropriate 50th percentile male anthropometric measurements, as defined in DOD-HDBK-743A, Military Handbook Anthropometry of U.S. Military Personnel, to allow for the necessary fit/seal that each piece of protective equipment requires.

3.2.3 The study will determine the feasibility of designing a PETMAN system whose materials of construction will not be significantly degraded by exposure to both traditional chemical agents (T) and Toxic Industrial Chemicals (TICs) / Toxic Industrial Materials (TIMs) (O) and that can subsequently be decontaminated to negligible levels without adversely affecting the operation of the PETMAN system as defined in 3.3.11.

3.2.6 The study will determine the feasibility of designing a PETMAN system capable of operating in fixed environmental chamber conditions (T) and a range of environmental chamber conditions (O) as defined in 3.3.6.1-3.3.6.5.

RELEVANT PETMAN REQUIREMENTS

PETMAN is conceived of as a surrogate for a soldier during the evaluation of an individual protection ensemble (IPE), specifically with respect to exposure to toxic industrial chemicals (TICs) and toxic industrial materials (TIMs). Just as a soldier would wear IPE, the PETMAN is expected to "wear" the IPE being tested; so it is important for the interaction between the PETMAN surface and the IPE to resemble the interaction between the soldier's skin and the IPE. At the same time, the surface must also provide the necessary protection for the mannequin during harsh testing conditions.

The relevant PETMAN requirements (Box 5.1) and potential architectures and candidate materials for the PETMAN skin are discussed below.

According to the PETMAN requirements, the mannequin surface must perform the following key functions:

1. Protect the internals—all the mechanical, electric, and computing modules that are housed inside the mannequin—under harsh test conditions. This includes exposure to liquid and vapor chemical agents and chemical simulants.
2. Simulate human skin so that the interaction between the IPE and the PETMAN surface resembles the surface interaction between the soldier's body and the IPE.
3. Facilitate the deployment of the sensors—preferably in a sensor network—for monitoring various parameters during testing.
4. Move with respiration to replicate the movement of the human chest wall with breathing.
5. Be easy and safe to decontaminate and, if necessary and feasible, be safely disposed.

Those key functions point to the need for a *skin* for PETMAN. It should be noted that the need for such a skin has not been explicitly stated in the Product Director, Test Equipment, Strategy and Support (PD TESS) requirements document.

THE PETMAN SKIN

The major characteristics of the PETMAN skin to meet the PETMAN requirements are breathability, sweatability, physiologic monitoring, resistance to chemical agents, usability, operating conditions, decontamination and disposability, and shape comformability.

Breathability defines the ability of the PETMAN skin to allow moisture vapor to escape from the body, whereas *sweatability* refers to its ability to

BOX 5.1
Relevant PETMAN Requirements

3.3.5 The PETMAN system shall be compatible with current under-ensemble chemical breakthrough sampling technologies, procedures, and equipment as defined in Test Operations Procedure (TOP) 10-2-022, Chemical Vapor and Aerosol System-Level Testing of Chemical/Biological Protective Suits (T) and designed to enable integration with real-time (1-second increments) sampling technologies, procedures, and equipment (O). At a minimum, sampling locations shall be the same as those defined in TOP 10-2-022.

3.3.6 The PETMAN system operation shall not be affected by the following chamber environmental conditions.
3.3.6.1 Temperature: 90°F ± 2°F (T); –25°F to 125°F ± 1°F, measured every 5 minutes (O)
3.3.6.2 Relative Humidity: 80% ± 3% (T); 0-100% ± 1%, measured every 5 minutes (O)
3.3.6.3 Wind speed: 0-10 mph ± 10% (T); 0-161 mph ± 2 mph (O)
3.3.6.4 Pressure: 0.25 iwg chamber vacuum maintained ± 2%
3.3.6.5 Liquid and vapor chemical agents including all nerve and vesicant agents, as well as the chemical simulants, triethylphosphate and methyl salicylate.

3.3.9 The PETMAN system shall be compatible with the individual protection and ancillary equipment listed in 3.3.9.1-3.3.9.11. The PETMAN system shall be designed such that the individual protection equipment can be properly donned IAW the respective technical manuals.

3.3.11 The PETMAN system shall be capable of being decontaminated with no adverse effects on the operation of the system and such that there is no effect on the next iteration of test (T) or leaving negligible agent residual, as defined by 3X decontamination level in DA PAM 385-61, Toxic Chemical Agent Safety Standards, (O) on the PETMAN system.

3.3.13 The PETMAN system shall record the following system parameters over time: skin temperature, respiration rate, perspiration rate, and total mass (in nanograms) of chemical vapor that penetrates/permeates through the protective ensemble. The PETMAN system shall record the start and stop time of each motion in 1-second increments.

NOTE: T= threshold and O=objective

transmit the moisture generated inside the PETMAN as human skin would. The skin should facilitate the integration of sensors to monitor physiologic measures, such as temperature and respiration rate, and to detect exposure to chemical agent, as well as TICs and TIMs. It should be *resistant to chemical agents* against which it is being tested. PETMAN should be *usable* in a variety of scenarios and with varied equipment, so its skin should not hinder handling of various weapons, moving around, or interaction with other components of the ensemble. It should withstand the various *operating conditions* of temperature, pressure, and environment to which PETMAN will be subjected to during the testing process. On completion of the test, the skin should be easy to *decontaminate* with such agents as bleach and hydrogen peroxide.[1] Finally, shape *conformability* is critical to the successful deployment of PETMAN: the skin should easily conform to cover the various contours and parts of PETMAN and to facilitate the movement of the chest wall with breathing.

With the need for a PETMAN skin established and its characteristics defined, its architecture must be defined and potential materials for creating it identified.

The skin architecture is the structure of the skin that needs to be designed to meet the defined PETMAN performance requirements. Figure 5.1 shows three conceptual skin architectures considered.

In the first architecture, the skin consists of one layer. It could be porous so that it "sweats," but it might be difficult to decontaminate if it is compromised during testing. Moreover, deploying sensors on the surface might be difficult because the sensors would need a mechanism for adhering to the surface; the sensors may also have to be powered, and they must communicate with the PETMAN sensor-controller module, necessitating the incorporation of wires or other devices.

In the second architecture, an outer skin is proposed to cover the PETMAN surface. This architecture overcomes the issues of decontamination (and disposal, if necessary) and deployment of sensors associated with the single-layer architecture. Through the proper selection of materials for the outer skin, it could be made soft to simulate the properties of human skin. However, if the soft skin and the surface skin are compromised, toxic agents could enter PETMAN and damage the internals.

To address the latter hazard, an impervious inner layer is proposed as shown in the third architecture. The third layer would be inside the PETMAN surface skin and provide an additional layer of protection to minimize contamination of the PETMAN internals.

Table 5.1 shows a comparative analysis of the three skin architectures

[1]DA PAM 385–61 Toxic Chemical Agent Safety Standards, 27 March 2002, Department of the Army Pamphlet 385-61, *http://www.army.mil/usapa/epubs/pdf/p385_61.pdf*, pp. 18-21.

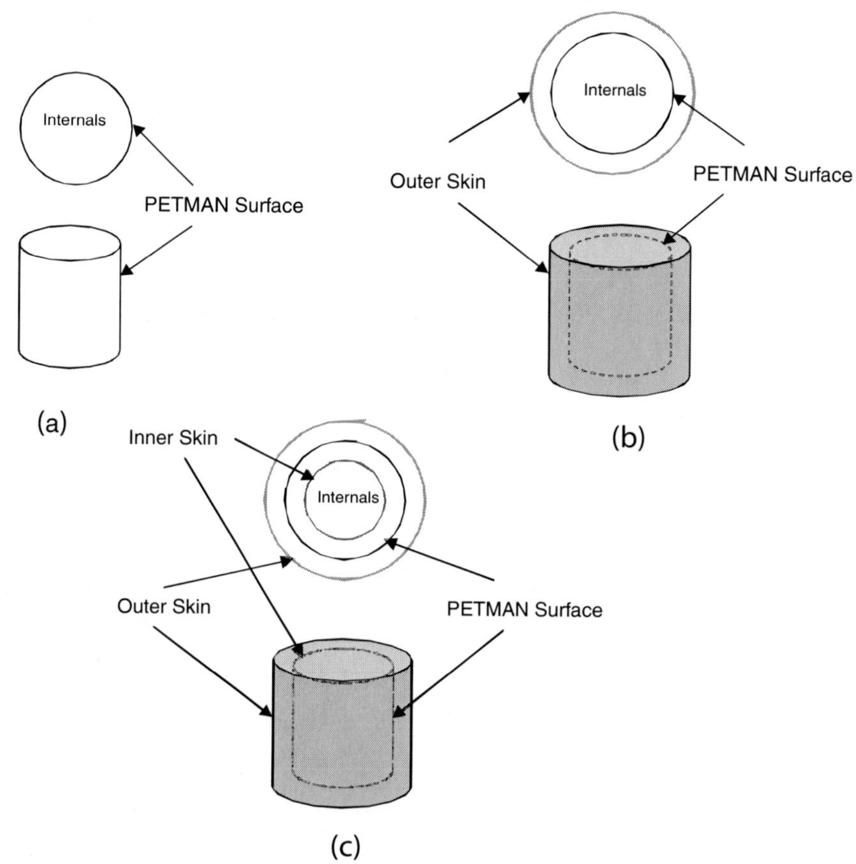

FIGURE 5.1 Conceptual architectures for the PETMAN skin: (a) one layer; (b) two layers; and (c) three layers.

based on a set of key evaluation criteria: degree of protection, ease of decontamination and disposal, ease of sensor deployment and reconfiguration, and manufacturability. A ranking scheme of H, M, and L (high, medium, and low) is used. As shown in the table, the one-layer architecture offers the lowest degree of protection and is difficult to decontaminate. The deployment of sensors on the surface is also difficult. However, from a manufacturabilty standpoint, it would be the easiest to fabricate.

The three-layer architecture offers a higher degree of protection than the two-layer architecture, but it would be more difficult to manufacture PETMAN in that architecture than in the two-layer architecture. By weighting the criteria (that is, using a weighted-priority matrix approach), the

TABLE 5.1 Comparison of Possible PETMAN Skin Architectures

Number of Layers	Degree of Protection	Ease of Decontamination and Disposal	Ease of Sensor Deployment and Reconfiguration of Positions	Manufacturability
One	L	L	L	H
Two	M	H	H	H
Three	H	H	H	L

NOTE: L = little or no capability, M = medium capability, H = highest capability or most desirable.

relative merits of the different architectures can be thoroughly evaluated, and such an approach is recommended during the design and development of PETMAN. In the present analysis, manufacturability is considered to affect the success of PETMAN substantially, so it is weighted more than degree of protection, which is another evaluation criterion.

CURRENT TECHNOLOGY TO MEET SKIN DESIGN CHALLENGES

On the basis of the comparative analysis, the preferred architecture of the PETMAN skin (of the three considered) would be two layers: a porous surface skin to protect the internals and an outer soft skin in which a sensor network can be easily deployed.

Surface Skin

The material for the surface skin must protect the PETMAN internals and facilitate sweating and zoned heating of the different regions. As discussed in Chapter 2, Measurement Technology Northwest (MTNW) has a metal porous skin (Figure 5.2) that has been used successfully in commercial mannequins for simulating sweating and providing zoned heating. It is a potential candidate for the PETMAN surface skin.

Soft Skin

In addition to having the characteristics of the PETMAN surface skin, the soft skin must facilitate the deployment of sensors, be easily decontaminated, and provide a surface like human skin for the IPE. An extensive review of the literature on artificial skin has led to the conclusion that developments in alternatives to skin are biological tissue-based and that the resulting structure typically requires a *living* body to sustain it and continue to grow. One of the best examples of such tissue-based artificial

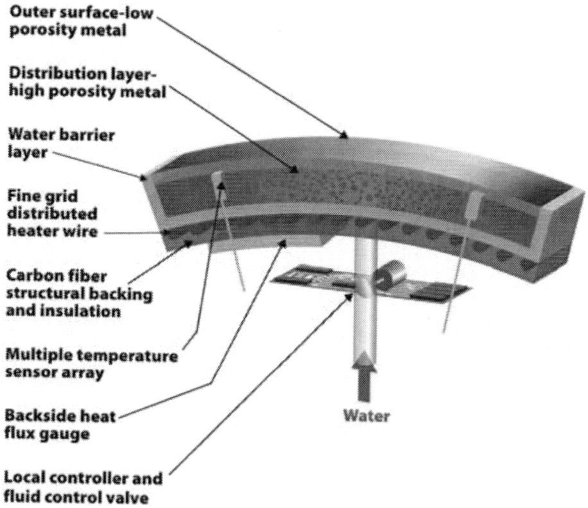

FIGURE 5.2 MTNW's metal porous skin, a potential candidate for PETMAN surface skin.
SOURCE: Rick Burke, Measurement Technology Northwest. *http://www.mtnw-usa.com*.

skins is EpiDerm™ from MatTek Corporation.[2] A conversation with a representative of the company led to the conclusion that no "polymer-based" surrogate could be an ideal substitute for human skin. Hence, the design challenge is to investigate and identify polymeric materials that would come close to human skin in physical, mechanical, and chemical properties and would meet the PETMAN performance requirements discussed earlier. During the course of the study, it was difficult to find published data on the physical and mechanical properties of human skin. Box 5.2 describes the structure and some properties of human skin.

The first step in selecting candidate materials for soft skin is to identify the specific properties (especially properties that can be measured in the laboratory) that will bear on the requirements. For example, breathability will be affected both by the material type and by the porosity of the structure used for the soft skin. The key design characteristics that must be considered in selecting materials for the soft skin include: porosity, flexural

[2]Epiderm Technical Specifications. MatTek Corporation. *http://www.mattek.com/pages/products/epiderm/specification*. Accessed June 15, 2007.

BOX 5.2
Human Skin: Structure and Properties

Human skin is a heterogeneous tissue composed of three superimposed layers that are intimately connected but distinct in their nature, structure, and properties (Figure 5.3).[a] The *epidermis*—mainly the *stratum corneum*—is concerned with protecting the organism from the environment. The fibrous *dermis* is a viscoelastic envelope that, with the *hypodermis* (the subcutaneous layer), plays an essential role in protecting the skin from mechanical stress. The mechanical function of the skin is the expression of the biomechanical nature of its components and their structural organization. Skin also keeps the human hydrated and cools humans with sweat.

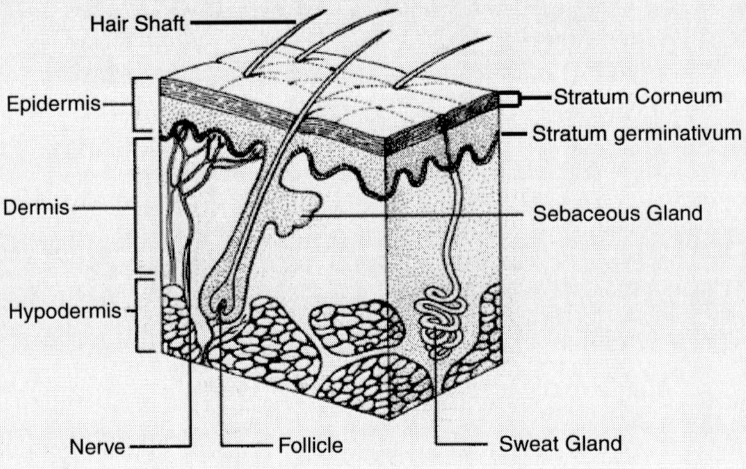

FIGURE 5.3 Structure of human skin.
SOURCE: Adapted from *http://nihseniorhealth.gov/skincancer/faq/faq3b_popup.html.*

---

[a]Escother, C., Rigal, J., Rochefort, A., Vasselet, R., Leveque, J., and Agache, P. G., "Age-related Mechanical Properties of Skin: An in vivo Study," *Journal of Investigative Dermatology*, Vol. 93, No. 3, 1989, pp. 353-357.

rigidity, abrasion resistance, manufacturability, form factor, surface hardness, chemical properties, and tensile properties.

The tensile properties and abrasion resistance of the material affect the durability of the soft skin, and its flexural rigidity influences shape conformability. Surface hardness will affect the interaction between PETMAN and the IPE; the chemical properties of the material define its resistance to TICs and TIMs and hence determine whether PETMAN requirements can be accomplished. The manufacturabilty of the material affects the eventual production of PETMAN. The form factor is the way in which the material can be used to create the soft skin, such as a coating, a casting, or a form-fitting layer (fabric or garment) on the PETMAN surface skin.

There are two approaches to achieving the PETMAN soft skin. The first is to use the design characteristics discussed above and engineer or create the materials for the soft skin. The second, more pragmatic approach, which has been adopted here, is to identify existing materials and carry out a comparative evaluation of their properties to select the materials that best meet the requirements. All the characteristics are important, but a subset (chemical properties, porosity, manufacturability, and form factor) are considered below in the evaluation process for selecting suitable materials for realizing the PETMAN soft skin.

Some potential candidate materials for the PETMAN soft skin are polyester, nylon, polyurethane, polyurethane in the form of spandex, expanded polytetrafluoroethylene (ePTFE) membranes and powder, and ePTFE-based fabrics. Some of the important properties of these materials are presented below.

Polyester: Polyester is made from polyethylene terephthalate (PET) and ariants, such as polytrimethylene terephthalate (PTT) and polybutylene terephthalate (PBT). Its tensile and abrasion properties are excellent, and it has high elastic recovery. Polyester has excellent resistance both to acids and alkalis, and it can be bleached with chlorine or oxygen bleach. It is used widely in traditional apparel and carpets.

Nylon: Nylon was the first synthetic polymeric fiber. It is a polyamide and can be made from hexamethylene diamine and adipic acid (nylon 6,6) or from caprolactam (nylon 6). It is strong and durable and has good elastic recovery properties. Nylon has excellent resistance to alkali and chlorine bleaches and is damaged by strong acids. It is used extensively in carpets, hosiery, and sportswear.

Polyurethane and spandex: Polyurethane offers the elasticity of rubber combined with the toughness and durability of metal. It can be manufactured in a variety of hardness, or durometers, and can elongate up to 800 percent and return to its original dimension without a substantial loss of memory.[3] Polyurethane is highly resistant to heat, cold, and aging. It has excellent long-term stability in water and is resistant to swelling and deterioration in temperatures as high as 80°C. It is resistant to oil and solvents and outperforms metal in chemical resistance. Those properties allow polyurethane to be used in some of the harshest of environments with minimal deterioration. It can also be bonded to metal, wood, or fabric.[4]

Spandex, known under the trade name Lycra®, which is made up of a long-chain polyglycol combined with a short diisocyanate, contains at least 85 percent polyurethane. It is known as an elastomer because it can be stretched to some degree and recoils when released. Because of its high stretch (600 percent), it is used in a wide variety of apparel products, especially form-fitting garments that take advantage of its unique elastic recovery properties, such as swimsuits, exercise gear, and undergarments.

Expanded polytetrafluoroethylene (ePTFE): PTFE is a polymerized tetrafluoroethylene, known for its chemical inertness, high thermal stability, low coefficient of friction, and other distinctive properties. When PTFE is stretched rapidly, it becomes a strong, water resistant yet breathable microporous material referred to as ePTFE, which is the key component of GORE-TEX® membrane.[5] The ePTFE structure is combined with an oleophobic, or oil-hating substance that allows vapor to pass through but prevents contaminating substances—such as body oils, cosmetics, insect repellents, and food substances—from penetrating. In addition, ePTFE is known to be chemically resistant to virtually all industrial chemicals, including acids, alcohols, aldehydes, amines, bases, esters, ethers, halogenated hydrocarbons, hydrocarbons, ketones, and polyalcohols.[6]

Another variant incorporating ePTFE, is the GORE CHEMPAK®, which combines a chemical protective polymer with an ePTFE. This fabric is also liquid-proof and air-impermeable and affords additional protection against liquid chemical-warfare agents and wind-driven agents in aerosol, vapor, and particulate form.

Table 5-2 shows a comparative assessment of these materials according to the major criteria identified earlier.

A key characteristic of PETMAN skin to meet the requirements is that it must be able to be decontaminated with the procedure outlined in

---

[3]Polyurethane. Bay Plastics. *http://www.bayplastics.co.uk/Product%20Materials/prod-polyurethane.htm*. Accessed June 15, 2007.

[4]Polyurethane—Features and Benefits. Elastochem Specialty Chemicals, Inc. *http://www.elastochem-ca.com/poly.html*. Accessed June 15, 2007.

[5]GORE-TEX. *www.goretex.com*. Accessed June 15, 2007.

[6]GORE Protective Vents Glossary. GORE. *http://www.gore.com/en_xx/products/venting/technical/membranevents_glossary.html#*. Accessed June 15, 2007.

TABLE 5-2 A Comparison of Potential Materials for PETMAN Soft Skin

Material	Chemical Resistant Properties	Breathability	Ease of Decontamination and Disposal	Surface Properties	Usability (Operating Conditions)
Polyester	M	L	M	M	L
Nylon	L	L	L	M	L
Polyurethane (including spandex)	H	M	H	H	H
ePTFE (including its variants)	H	H	H	H	H

NOTE: L = little or no capability, M = medium capability, H = highest capability or most desirable.

Toxic Chemical Agent Safety Standards.[7] Polyurethane and ePTFE are far superior to polyester and nylon in chemical resistant properties, for which they are rated high in the table. Nylon is affected by strong acids and so is rated low. Polyurethane and ePTFE are both highly inert chemically and are assigned a rating of high to denote the ease with which they can be decontaminated. Polyester can be decontaminated with both chlorine and oxygen bleaches, whereas only chlorine bleach can be used on nylon. The surface characteristics of polyurethane and ePTFE are superior to those of nylon and polyester. Similarly, the ability of polyurethane and ePTFE to withstand extreme operating conditions (such as those to which PETMAN will be subjected) merits the ratings shown in the table in comparison with those for polyester and nylon.

Form Factor for PETMAN Soft Skin

The PETMAN soft skin can take one of three forms: It can be a coating on the surface skin, for example, using polyurethane[8]; it can be a casting that is bonded to the surface skin with a breathable adhesive[9]; or it can be a form-fitting layer on the surface skin in the form of a fabric or garment, as is the case with NEWTON, the mannequin from MTNW.[10] A comparative evaluation of the three options based on a set of critical criteria is shown in Table 5.3.

From the viewpoint of ease of sensor deployment, a form-fitting layer provides the greatest ease because the sensors could potentially be integrated into the structure. In contrast, deploying the sensors (especially providing power and communication) would not be easy on smooth surfaces realized by coating and casting. A form-fitting layer is also better than a coating or casting when it comes to ease of decontamination and disposal.

A surface coating or casting would not affect the movement of PETMAN around the joints in the legs, arms, and so on. In comparison, a form-fitting layer, if not properly designed, potentially could impair the movement in the joints; and the form-fitting layer could be subjected to deformation from repeated bending and flexure. With respect to facilitating chest-wall movement, the form-fitting layer would be the most accommodating. Finally, from a manufacturability standpoint, a form-fitting layer would be the easiest to manufacture and also the most economical. Thus,

[7]U.S. Army. DA PAM 385–61 Toxic Chemical Agent Safety Standards, 27 March 2002, Department of the Army Pamphlet 385-61, *http://www.army.mil/usapa/epubs/pdf/p385_61.pdf*, pp. 18-21.

[8]Sakhpara, U.S. Patent 4,942,214, 1990.

[9]Driskill U.S. Patent 4,925,732, 1990.

[10]Burke, R., Presentation at the 3rd PETMAN Meeting, April 2007 (see Appendix D).

TABLE 5.3 Comparison of Possible Material Form Factor for PETMAN Soft Skin

Form Factor	Ease of Sensor Deployment and Reconfiguration of Positions	Ease of Decontamination and Disposal	PETMAN Joints (Arms, Legs, and so on)	Chest-Wall Movement	Manufacturability
Coating	L	L	H	M	M
Casting (bond with breathable adhesive)	L	M	H	L	M
Form-fitting layer	H	H	M	H	H

NOTE: L = little or no capability, M = medium capability, H = highest capability or most desirable.

on the basis of this structured evaluation, a potential form factor for the PETMAN soft skin is a form-fitting layer.

Architecture of PETMAN Soft Skin

The final step is to identify a potential configuration of material type *and* form factor for the PETMAN soft Skin. Table 5.4 shows the composite ranking of the various combinations of form factor—casting and a form-fitting layer—and the five choices for materials, including ePTFE and its variants, and polyurethane and its variant spandex.

With the powdered form of ePTFE,[11] it is possible to produce a casting, but use of ePTFE in a form-fitting membrane would be preferable. In Table 5-3, a form-fitting layer was the preferred form factor. The ePTFE-based CHEMPAK fabrics are available only in fabric form, so they cannot be used for creating a casting. Polyurethane can be easily cast compared with its use as a form-fitting layer). Spandex is ideally suited for use in a form-fitting layer (compared with casting). A form-fitting layer has the additional advantage of serving as the infrastructure for the deployment of sensors, potentially using the wearable-motherboard paradigm.

In Table 5-2, ePTFE was identified as the preferred material over polyurethane. Therefore, on the basis of both form factor and properties of materials, a form-fitting layer of ePTFE-based materials is the preferred configuration for the PETMAN soft skin. Because the CHEMPAK fabrics have additional chemical-protective characteristics, they could be an ideal choice for the PETMAN soft skin. Alternatively, a spandex-based form-fitting layer could serve as another choice for soft skin.

Integrating the PETMAN Sensing System and Skin

The next step in the process of designing a PETMAN will be to identify an effective means of integrating the skin into the PETMAN system. In particular, it is important to consider integrating the skin and sensors identified in Chapter 3. Analysis of the requirements leads to the following conclusions:

- Different types of sensors are needed to monitor different characteristics.
- Different numbers of sensors of each type may be needed for computing a single characteristic
- The sensors need to be positioned in different locations on PETMAN to acquire the proper signals.

[11]Dolan et al., USP 5,646,192, 1997.

TABLE 5.4 Soft-Skin Configuration: Form Factor and Materials Analysis

Form Factor	Material Options				
	ePTFE and Variants	CHEMPAK® Ultra Barrier	CHEMPAK® Selectively Permeable	Polyurethane	Spandex
Casting	M	N/A	N/A	M	L
Form-fitting layer	H	H	H	L	H

NOTE: L = little or no capability, M = medium capability, H = highest capability or most desirable.

- Different subsets of sensors may be used at different times, necessitating easy attachment and removal.
- The sensors and the manner in which they are deployed should neither impair the mobility of the PETMAN nor affect the interaction between PETMAN and the IPE.
- Sensors may need to be powered.
- Sensors need to communicate with the PETMAN sensor control module.

Thus, what is needed is the design and implementation of a network of sensors on PETMAN to meet the threshold and objective requirements. Smart textiles or wearable electronic systems provide a possible platform for such a network.[12] For example, the Smart Shirt uses optical fibers to detect physiological signals and movement.[13]

Figure 5.4 shows the architecture of the imbedded sensor network. The base fabric provides the necessary physical infrastructure and is made of typical textile fibers chosen according to the intended application. The optical fiber integrated into the structure provides the infrastructure for carrying information through the fabric and is used for identifying projectile penetration with optical time-domain reflectometry. The interconnection technology has been used to create a flexible conductive framework to plug in sensors for monitoring a variety of vital signs. This technology can potentially be adopted for deploying sensors on PETMAN; the optical fibers in the fabric can serve as the infrastructure for the fiber-optics-based sensors, and the conductive-fiber network can provide the power and communication capabilities required by the other types of sensors on PETMAN.

CONCLUSIONS AND RECOMMENDATIONS

Analysis of the PETMAN surface requirements and potential solutions to meet the requirements resulted in the following conclusions and recommendations:

> **Conclusion 5-1:** The need to simulate the interaction of human skin with the IPE has not been specified in the PETMAN requirements. However, since the PETMAN is conceived of as a surrogate for a soldier during the evaluation of IPE, it is important for the interaction

[12] Service, R. F. 2003. News Focus Technology: Electronic Textiles Charge Ahead. *Science* 301(5635):909-911; Park, S., and S. Jayaraman. 2003. Smart Textiles: Wearable Electronic Systems. *MRS Bulletin* (August 2003):585-591.

[13] See *http://www.sensatex.com/index.php/smartshirt-system* Accessed August 16, 2007.

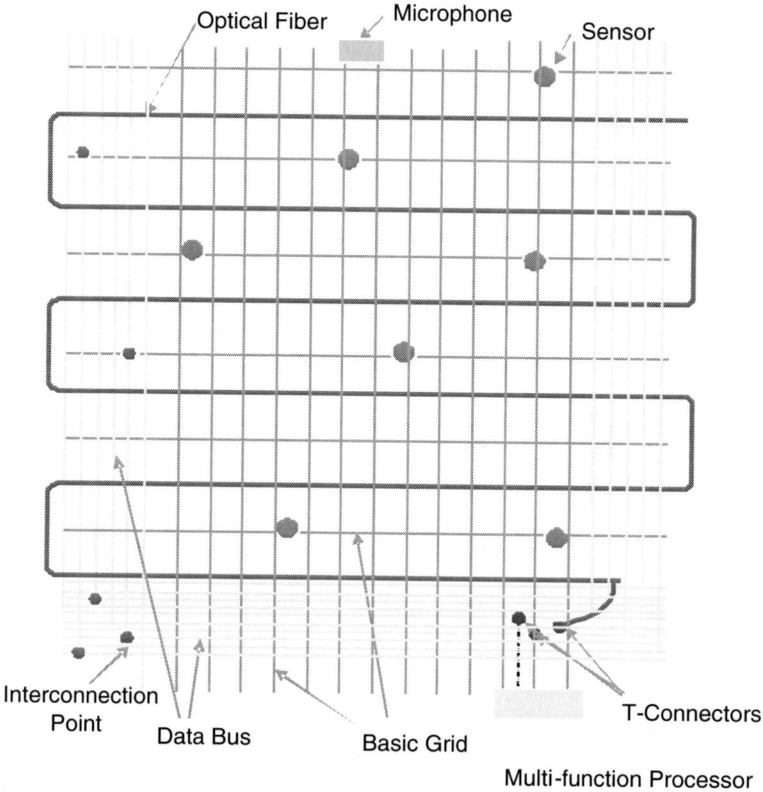

FIGURE 5.4 The architecture of an embedded sensor network.
SOURCE: S. Jayaraman, Georgia Institute of Technology.

between the PETMAN surface and the IPE to resemble the interaction between the soldier's skin and the IPE.

Recommendation 5-1: Simulation of the interaction of human skin with the IPE should be added as an objective PETMAN requirement.

Conclusion 5-2: Various methods exist for achieving the threshold and objective PETMAN requirements that include a mannequin surface consisting of multiple layers of skin, such as an inner skin that protects internal mechanical and electrical components and an outer soft skin that is in contact with the IPE.

Conclusion 5-3: Various methods exist for the deployment of sensors on PETMAN, such as sensor-embedded textiles.

Recommendation 5-2: There should be a multilayer skin architecture for PETMAN to meet the threshold and objective requirements, such as an inner porous skin and an outer soft skin with an embedded sensor network.

Conclusion 5-4: Various materials exist for the inner skin to meet the threshold and objective requirements of PETMAN, such as a metal porous skin from Measurement Technology Northwest that perspires and provides zoned heating.

Conclusion 5-5: Various materials exist for the outer soft skin in contact with the IPE, such as expanded polytetrafluoroethylene and polyurethane (spandex).

Conclusion 5-6: Extensive data on the performance characteristics of the metal porous skin and ePTFE-based materials were not available, especially those related to the chemical and biologic protection characteristics of CHEMPAK.

Recommendation 5-4: The performance characteristics and metrics claimed by the commercial developers (vendors) of metal porous skin and ePTFE-based fabrics and materials should be validated as part of the PETMAN design and development process.

6

Design Challenge: An Integrated PETMAN System

This chapter addresses overall integration of the PETMAN system as described in section 3.2 of the requirements document (Appendix B):

The feasibility study will be based on the PETMAN system performance requirements defined in 3.3.1-3.3.14. The focus of the feasibility study will be on the significant design challenges defined in 3.2.1-3.2.8 of the PETMAN system design in the feasibility study, however all requirements must be considered while conducting the study. The feasibility study will determine the requirements trade-offs and cost differentials associated with each design approach considered in the feasibility study.

PETMAN SYSTEM DESIGN OVERVIEW

PETMAN will be a complex system consisting of many functional subsystems required to perform a variety of behavioral and sensing tasks. A discussion of full system integration must address physical, communication, and control integration. A typical engineering systems approach to this problem will create an overall systems architecture design that identifies the specific submodules and their functions and interrelations. The engineering systems approach is important to modularize the design and evaluation process and to provide a systematic framework for the development process. Principal engineering functions—such as sensing, actuation, modeling, and calibration—are identified, and the analytic design required is focused on those elements. In addition, the engineering systems view provides a basis for the information system architecture (both hardware and software). For PETMAN, the three main kinds of integration that must be considered are

- **Physical integration,** which deals with packaging of all the components deemed appropriate, available, and supportive of desired performance levels.
- **Controls integration,** which addresses the complexity of the multiple subsystems that make up PETMAN and their interdependence.
- **Communication,** which addresses the smooth and seamless exchanges between the subsystems and external monitoring stations to attain desired performance.

The feasibility of full system integration depends on the degree of complexity of PETMAN's physical subsystems, that is, the number of actuators, the number of sensors, the mode of communication between system components, and performance demands with respect to power, endurance, and complexity of physical tasks (exercises). Integration also depends on the features of the preferred design characteristics of the subsystems, such as skin construction and physical characteristics, wired or wireless communication, and modular or discrete assemblies.

Figure 6-1 is a conceptual diagram of a PETMAN systems design. The system is organized as a hierarchy of subsystems associated with the major functions.

Task Planner

The task planner provides the overall coordination of the system and designates the main system behaviors that are required and their key characteristics, including timing, rate of motion, sensitivity of sensors, and type of protective ensemble. The task planner also supports the principal user interface and access to software systems.

Sensor Subsystem

The sensor subsystem supports the array of sensors (as described in Chapter 3) that detect chemical signatures in the space between the protective ensemble and the skin. A sensor planning module is responsible for planning and monitoring the execution of the sensor functions. A given application may require a particular subset (by location or type) of sensors, and the sensor planning module selects, calibrates, and confirms the sensor requirements. The sensor planning module also selects the preprocessing and postprocessing required for management of the sensor data acquired. The sensor data themselves are received by a sensor control and process-

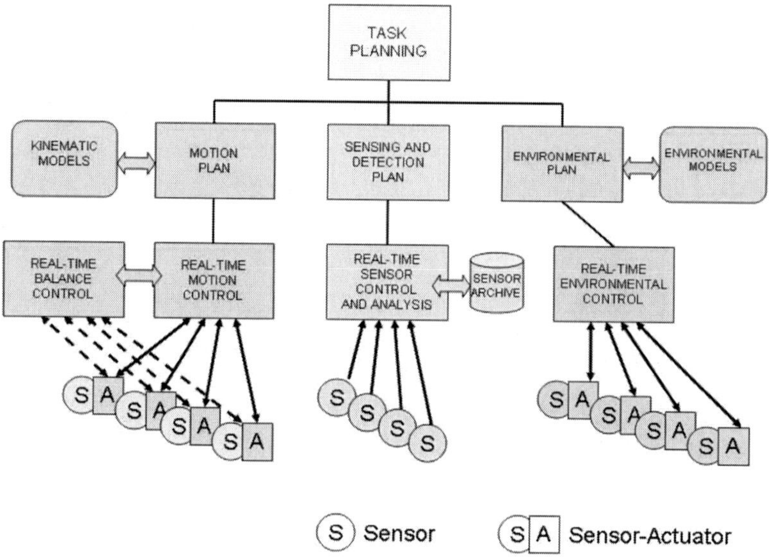

FIGURE 6.1 Conceptual design of the integrated PETMAN system.

ing module that has direct interfaces with the sensor elements. The sensor control and processing module could be either a centralized or a distributed information system. Data received and preprocessed are stored in the sensor data archive (either centralized or distributed and either onboard or remote). Retrieval of stored data is managed by the sensor planning module.

Environmental (Under-Ensemble) Subsytem

The environmental subsystem determines and controls the conditions in the protective ensemble—understood to be principally air temperature and humidity. The environmental planning module determines these required conditions, including the mode of control that is intended for the subsystem in various portions of the ensemble-skin air space. For example, one mode might require a set skin temperature and a set perspiration (water-transport) rate at the surface of the skin, and another mode might estimate the activity of the body and use a metabolic model to control air temperature and humidity in accord with simulated body metabolism; the second mode would require more extensive control systems and more elaborate models to govern heat and water-vapor release. The actuators for the environmental system would probably be embedded at or beneath the skin surface.

Motion Subsystem

The motion subsystem would provide for the control of all motions of limbs, head, and chest associated with the performance requirements. A critical assumption regarding the performance requirements for PETMAN is that the motion requirements can all be predetermined for each phase of the study. That is, each arm motion, walking motion, or other maneuver is preprogrammed and noninteractive with unknown constraints or events in the local environment. That assumption is critical to the systems design and greatly simplifies the requirements. If all motions under known conditions are preprogrammed, many of the kinematic and dynamic control requirements can be precomputed, and execution is greatly simplified. The one exception to the assumption may be related to balance. If true dynamic balance is required (that is, no external stabilizing elements or other mechanisms are used), true dynamic control is required to support the system, and both software and hardware requirements (for example, joint sensing and control sample rates) are increased.

The motion planning module would interpret motion behaviors (such as walking) set by the task planner and translate them into elemental limb and joint motions (for example, moving the knee) to achieve complex coordinate motion. The overall degrees of freedom (DoFs) of the mannequin are not known, but one might expect 16-48 DoFs that must be specified (by the motion planner) and controlled (by the motion controller). Such coordinated control typically requires a comprehensive kinematic model of the system that would constrain the motions in an integrated fashion. The introduction of dynamic balance would require still further modeling of system dynamics.

The motion control module would integrate joint and limb sensors with command signals to joint and limb actuators to execute the motion plan. Specific control algorithms for motion and balance could be derived with a variety of analytic, learning, or heuristic approaches. Such approaches have been extensively studied. For dynamic motions (such as jumping), it is less clear that performance could be achieved without more extensive development.

Several specific issues arise in relation to the motion subsystem. The use of anthropomorphic hands for general tasks is not well understood, and assumptions about manipulating objects to mimic normal human motions may not be realistic. The choice of hand design and the sensor suite available (for example, tactile and force sensors) will be important in determining the range of manipulation motions and the tasks that are feasible. A second open question may be related to chest-wall motion. This motion might be coupled to transport of air to the respirator, but that is not necessary for the functional design. Chest-wall motion might be achieved through separate mechanisms and coordinated through control functions with air transport.

Taken independently, it is technically feasible to construct almost all the PETMAN subsystems with some degree of compatibility with individual threshold or objective performance requirements. However, as requirements are grouped, power requirements increase, weight is increased, more packaging volume is needed, higher control processing power is required, and more complex controls and communication schemes are needed. On the basis of current state of the art of mannequins and humanoid robots, an autonomous nontethered mannequin that performs at the objective level is not feasible; it would entail the sacrifice of some critical performance requirements.

That assessment is based on a comparison of the autonomous operation attainable with state-of-the-art technology of portable energy plants and their charging frequency, power rating, and weight requirements.

In contrast, the feasibility of approaching more of the performance demands, especially of performing exercises and carrying weapons at desired rates and endurance, increases appreciably when power is supplied externally, that is, with a tether. Once a tether is accommodated for power, control and communication can be appreciably simplified because higher power controls can be communicated with externally.

Accordingly, we give special attention here to how performance requirements may be met to help in the decision of whether to tether or not to tether.

TETHERING COMPATIBILITY WITH FUNCTIONAL REQUIREMENTS

Table 6.1 provides some insight into the compatibility of tethering with critical functional requirements at the objective and threshold levels. Compatibility is given a qualitative assessment that indicates whether tethering or nontethering low (L), medium (M), or high (H) compatibility with the desired or required performance. Requirements are listed in the table by paragraph number of the PETMAN proposal (see Appendix B) with key identifying statements.

Tethering would be required to supply locomotive power for the mannequin's joint motors, to supply fluids necessary for simulating biologic functions, and to communicate control signals and data to and from different sensors and actuators. The concept of tethering may also be expanded to include external manipulation of the mannequin's limbs if self-actuation proves too demanding within the constraints of objective or threshold performance levels.

The individual performance requirements of PETMAN compete for

TABLE 6.1 Comparative Assessment of Capability of a Tethered and a Nontethered System in Meeting the PETMAN Functional Requirements

Design Challenges and Functional Requirements		Tethered	Non-Tethered
3.2	**Feasibility Study Design Challenges**		
3.2.1	Integrity of individual protection ensemble equipment	M	H
3.2.2	Compatible with individual protection, ancillary equipment, and weapon systems	M	H
	• Donning and doffing	M	M
	• Meets 50th percentile males	H	L
3.2.3	Materials of construction not degraded by	M	M
	chemicals	M	M
	• Can be decontaminated		
3.2.4	Integrated into ensemble sampling system	H	M
3.2.5	Control of skin temperature and perspiration and respiration rates	H	M
3.2.6	Not affected by specified environmental chamber conditions	H	M
3.2.7	a- Man-in-Simulant Test (MIST) exercises b- System performing human-like movements	See 3.3.8 below	See 3.3.8 below
8.8.8	Fully articulated hands and feet that simulate human motion	L	L
3.3	**Performance Requirements**		
3.3.2	Operation for (T/O):		
	a- (12/24) h before operational maintenance	M	L
	b- (3/6) months before preventive maintenance	M	L
	c- (6/12) months before calibration	M	M
3.3.3	Anthropometric requirements; 50th percentile male	M	L
3.3.4.1	Skin temperature fixed/variable	M	M
3.3.4.2	Perspiration rate fixed/variable	M	M
3.3.4.3	Respiration rate fixed/variable	M	M
3.3.5	Compatible with chemical-breakthrough sampling technologies	H	M
3.3.6	Not affected by chamber environmental conditions: 3.3.6.1, 3.3.6.2, 3.3.6.3 (for 0-10	M	L
	mph), 3.3.6.4, 3.3.6.5	L	L
3.3.6.3	Not affected by wind speed up to 161 mph		
3.3.7	Articulated and robotic, like a human in aesthetics and proportions:	H	L
3.3.7.1-9, -11, -12, -13, -14, -15, -16:	arms, legs,	H	M
	torso	L	L
3.3.7.10	hands		

TABLE 6.1 Continued

Design Challenges and Functional Requirements	Tethered	Non-Tethered
3.3.8 Simulation of exercises:		
3.3.8.1 Standing	M	M
3.3.8.2 Walking at 4.8 km/h (3 mph)	H	M
3.3.8.3 Marching (12-in.-high step) at 4.8 km/h	H	M
3.3.8.4 Jumping jacks—jumping, landing	M	L
3.3.8.5 Sitting	H	M
3.3.8.6 Standing to squatting to standing position	M	L
3.3.8.7 Reaching arms in all directions	M	M
3.3.8.8 From standing to lying prone and return	M	L
3.3.8.9 Kneeling on one knee	M	M
3.3.8.10 Kneeling on both knees	M	M
3.3.8.11 Low crawl	M	L
3.3.8.12 High crawl	M	L
3.3.8.13 Aiming weapon in various positions	M	L
3.3.9 Compatible with protection and ancillary equipment:	M	M
3.3.9.1 Suits	M	L
3.3.9.2 Boots	M	M
3.3.9.3 Gloves	M	M
3.3.9.4 Masks	M	M
3.3.9.5 Helmets	M	L
3.3.9.6 Combat boots	M	M
3.3.9.7 Ballistic-protection vests	M	M
3.3.9.8 Pistol holsters	M	M
3.3.9.9 Battle dress and combat uniform	M	M
3.3.9.10 Physical training gear, T-shirt, shorts, and so on	M	M
3.3.9.11 Skin-exposure reduction paste		
3.3.10 Compatible with weapons systems:		
3.3.10.1 M4 modular weapon	H	M
3.3.10.2 M24 sniper rifle	H	M
3.3.10.3 M16A2 rifle, 5.56 mm	H	M
3.3.10.4 XM8 lightweight assault rifle	H	M
3.3.11 Capable of being decontaminated	M	M
3.3.12 Uses common commercially available parts	H	M
3.3.13 Records system measures at 1-s intervals:		
Skin temperature	M	M
Respiration rate	M	M
Perspiration rate	M	M
Penetrating mass (in nanograms) of chemical vapor	M	M
3.3.14 Programmable to perform exercises or motion	H	M

NOTE: H = high capability, M = adequate or medium capability, L = little or no capability.

power resources and controllability and are often in conflict. For example, extensive articulation is required for human-like aesthetics, but when coupled with required exercises with real weapons the weight and size would probably become too great to meet the 50th percentile of male subjects. Similarly, increased weight requires higher power, which conflicts with long performance times in nontethered systems.

Accordingly, the assessment of compatibility depends on the priorities associated with each of the PETMAN performance requirements. Determining such priorities would move the PETMAN system into the realm of meeting more objective level requirements.

The assessment is highly subjective, but it is indicative of the potential opportunity for meeting the performance requirements when either approach is adopted. Variations in judgments are possible, although the overall assessment is likely to be the same.

The overall systems design of a PETMAN is strongly influenced by the choice of a nontethered or tethered operation. With the nontethered approach, all resources—including those required for power, gases, liquids, and heating and cooling functions—would need to be housed and managed inside the mannequin itself. The principal advantage of nontethered operation is that the ensemble remains entirely intact in accord with the goal of using standard ensemble equipment without modification as it would be worn by a human. Any tether access to the mannequin would need to penetrate the ensemble at some point or require customization of some part of it to enable access. As described below, a tether connected to the head, torso, or boot heel of the mannequin may constitute an acceptable compromise.

A review of available technologies suggests that fully self-contained nontethered operation may be difficult to achieve in the short term because of the space and weight requirements associated with onboard power and materials. Current battery technologies will realistically support electric motor drives only for self locomotion of the mannequin with simple movements. Similarly, hydraulic systems pose major challenges to the nontethered design in that volume and power for an onboard hydraulic pump, storage tank, accumulator, and necessary battery power source are unrealistic with current technologies. On the basis of those assessments, it is important to consider the principal options for tethered operations and to identify their effects and risks. The key favorable and unfavorable aspects of the use of a tether include the following:

Favorable:

- Provides access for ample power, gases, liquids, and communication.
- Dramatically reduces the space and weight requirements for inter-

nal components of the mannequin, making the 50th percentile male envelope more realizable.
- Might make it possible to operate for the targeted operational periods and maintenance frequencies.
- Allows most of the functional requirements to be met.

Unfavorable:
- Requires some modification of the ensemble to enable penetration by a tether that contains cables, tubes, and fibers.
- May disturb the natural mobility, dynamics, and range of motion for the mannequin because of the forces and weight of the tether itself.
- May require specially actuated support capability to manipulate the tether as the mannequin moves.

On the other hand, the key favorable and unfavorable aspects of the nontethered approach include the following:

Favorable:
- Maintains the integrity of the ensemble for testing without penetration.
- Provides good aesthetics.

Unfavorable:
- Has inadequate power to meet many of the motion and actuation requirements for full mannequin functionality.
- If required power is provided internally, weight and size are likely to far exceed those of the 50th percentile male.
- Is not expected to meet targeted operational periods and frequency of maintenance if the 50th percentile male envelope is observed.
- Posture stability is not yet controllable for most of the exercise movements.

Tether Attachment Options

Three broad options for tether attachment to the mannequin are discussed below.

Head

a. *Back of head.* Tether connection to the back of the head would require modification of the helmet and all hood components of the ensemble.
b. *Face.* Integration of the tether connection into the face mask may be easier to accommodate as an equipment modification, but there may be disruption of normal sealing of the mask because of forces applied by the tether.
c. *Both.* Back and face tethers could interfere seriously with head and neck motion. Without active suspension of the cable, a balancing mannequin would be affected by such forces applied to the head. Abnormal motions might affect penetration in the head and neck regions.

Torso

For the back or front of the torso, the principal areas of every test ensemble would need to be modified to introduce a tether interface. Tethers attached to the torso could also interfere substantially with motion of the mannequin because of applied forces, and active manipulation of the tether might be needed.

Foot or Boot

Tether attachment to the sole or heel of the boot may have some advantages, especially that any penetration may be limited to one shoe or boot of a pair, allowing full testing opportunity for the other shoe or boot of the same pair.

- The modification of a boot may be more acceptable than modification of other components of the ensemble, particularly because the sole and heel of the boot may be less susceptible to the leakage under study. In addition, it may be more acceptable to use a small number of custom boot models that are customized with tether connectors, and these choices may have minimal effects on other aspects of the tests.
- Some forms of attachment of the tether to the sole of a boot could directly interfere with foot placement for all standing and walking motions. Such attachments may lead to an alternative approach to generate motion externally; for example, each foot would be moved by an external manipulation mechanism without actual ground contact in achieving the motion. Such an approach would lead to a quite different overall systems design (external-internal

drive), considered below, but might accommodate the boot-based tether connection.

- The attachment of the tether to the back of the heel of the boot may be acceptable if a small number of modified boots could be used in a variety of tests. Attachment to the heel would still permit natural motion of the foot and boot and would not interfere with standing, walking, jumping, or crawling motions. External manipulation of the tether to avoid applied forces to the foot and leg may be necessary.

External-Internal Drive of the Mannequin

The self-contained (nontethered) mannequin design would require all actuators to be inside the body, and this poses a challenging design problem to fit actuators into a body of the required size and the problems of managing power and cooling to support the actuators. An alternative strategy could configure external manipulation of the mannequin limbs with external mechanisms. A manipulator with four to six DOFs might be attached to each foot and to each hand. The external drives would provide the primary forces used to move the feet and hands and might, in principle, provide much of the balance and dynamic support needed for the set of prescribed motions. The mannequin would still require some internal drive mechanisms, but the power, DoFs, and range of motion for the powered actuators would be reduced. In addition, passive joints could be introduced into the mannequin to support the needed DoFs. In principle, the resulting system could be used to generate the required motions. There would be a loss of realism in the lack of actual boot contact with the ground. The execution of walking, marching, jumping, and crawling motions could still be accomplished to mimic the required motions.

This approach may have merit for nontethered mannequins for the performance of tasks and exercises that require higher loads and speeds than may be possible with internally packaged actuators, such as crawling with weapons.

SOME INTEGRATION OBSERVATIONS

The design of the PETMAN system—mannequin, controls, software, subsystems, and the rest—is challenging. However, all the individual subsystems appear manageable independently, and the challenge is in the integration within the constraints of size, weight, and functional requirements. Accordingly, a systematic approach to the design that focuses on attaining the most functionality at the early stages and prepares the design to accommodate remaining functionality at later stages is desirable.

The nontethered mannequin could be considered as the ultimate objective, but it should not now be the overriding one. It appears advantageous to begin with a tethered approach and a target of meeting threshold performance levels. The design process would determine the configuration and characteristics of the subsystems and system components, and it should be unencumbered by power requirements for locomotion or control. Once the threshold levels are achieved, higher-priority objectives could be incorporated progressively to the extent that volume and weight restrictions allow. A follow-up phase could deal with progressive enhancements as technology is developed and experience leads to more ingenious designs and inventions.

There is a high level of uncertainty in the feasibility of attaining many of the PETMAN requirements. That uncertainty makes it impossible to predict the cost and time requirements of a PETMAN-like mannequin solicitation. A phased approach may be useful, with the initial phase being a development phase that addresses the high-priority requirements. That phase should result in a realistic set of specifications that address those requirements, and focus on objectives that can be attained within the desired timeframe.

A minimum of one year may be necessary for the development phase. The follow-up phase may then focus on the realization of a mannequin with acceptable capabilities. The following are areas to be addressed in a development phase:

- Respiration volume and space requirements.
- Actuation of hands and feet within the volume available in the arms and legs of the 50th percentile torso.
- Stability and control of the high-activity exercises.
- Decontamination.
- Level of functionality without a tether.
- Heating and cooling.
- Maintainablity and reliability.

PRACTICAL APPROACH TO MANNEQUIN DEVELOPMENT

A practical approach to developing the PETMAN system may follow these steps:

1. Design the mannequin so that all actuators and controls are externally powered and a tether penetrates at the least disturbing location, possibly one of the boots.
2. Operate the electric motors at powers higher than their continuous-operation ratings and improve their duty cycle by cooling supplied

through the tether. This approach reduces the actuators to their smallest practical size.

3. Consider using the highest-speed motors with higher-ratio speed reducers, harmonic drives, and other cycloidal reducers and allowing compact packaging.
4. Assess whether the actuators and other subsystem components can be packaged within the target volume and size constraints of the 50th percentile male population. If not, apply acceptable compromises and trade-offs to reduce the number of actuators or their power or size requirements to acceptable and packageable levels; some low-priority functionality could be sacrificed, beginning with the ones that are most demanding of power and space.
5. Reevaluate the possibilities of improving functionality and regaining compromise losses by replacing some electric actuators with air actuators, especially those with small stroke-angle movements; consider trade-offs to use oversized motors rather than motor-reducer combinations.
6. Reassess the tethering of controls and communication versus self-contained approaches and wireless communication to reduce the size, rigidity, and weight of the tether and help with its maneuverability and management.
7. Allow for efficient use and distribution of cooling air to reduce its supply-line size; for example, if not all actuators are cooled simultaneously, cooling air may be cascaded from one function to another for cooling particular areas or heating others as needed to maintain body-surface areas at desired temperatures; consider circulating heated air with blowers and fans.
8. A whole-system energy distribution study may be performed to help to distribute the cooling and heating media most efficiently. Ideally, if the net balance is an energy surplus, only cooling is required, although in reality some heating will be necessary in some locations and more cooling will be added; such inefficiency should be minimized.

SYSTEMS ARCHITECTURE AND SOFTWARE CONSIDERATIONS

The architecture for all the systems reviewed takes the form of a hierarchic structure that integrates task definition and planning, model-based execution and representation, and behavior-based sensor reflex control. This type of architecture is generally well understood, the implementation is feasible, and such a modular approach is important for the reliable and systematic development of PETMAN. The dancing humanoid by Ikeuchi's group is an example of behavior control similar to what is needed for the PETMAN project (see *www.cvl.iis.u-tokyo.ac.jp*).

Software systems for PETMAN must support several classes of users to meet the stated needs fully. These systems include

- *Operations testing and monitoring.* This would be a user-oriented system interface designed for nontechnical users and prepared with clear interactions for routine setup and running of tests, collection and preview of data, and periodic general monitoring as required.
- *Test development and routing maintenance.* This system would support periodic revision and development of new test procedures within the basic framework of the PETMAN system. It would integrate simulation and modeling capabilities so that a user could create new exercises and behaviors and visualize them in the system design mode. No formal programming language interaction should be required.
- *Expert programming.* This system would provide expert access to the modular computer software codes that underlie the hierarchic system and would be used by experts, perhaps vendor staff, trained to work with it at this level.

Existing software tools and systems are available to create all those components. In addition, it is recommended that state-of-the-art software engineering methods and procedures be instituted to support the creation of reliable and maintainable software.

The software structure of PETMAN will be critical for both the functionality and the usability of the system. As suggested in the system architecture overview discussed earlier, the system will most likely be hierarchic in structure, and the software should be well organized, modular, and clearly structured to implement key capabilities for planning, modeling, control, sensing, and actuation. In practice, it may be appropriate to use different programming environments and languages for the different layers of software. Although the planning and task-development environment might use an abstract representation to define sequences of mannequin behaviors, the low-level embedded software that drives motors and sensors will involve a quite different style of programming and implementation. From a broad systems perspective, software development itself is feasible with existing software methods and programming environments. However, demonstrated experience in development at all three levels of capability should be examined carefully. Clear benchmarks for development will be critical and require adequate effort and resources.

A key approach to programming takes into account the variety of users of the system. At least three groups of users must have effective and efficient access to appropriate software tools in the final implemented systems, as described below.

Real-Time Test Implementation and Monitoring

This level of software development should focus on user interaction and capabilities for users to implement and monitor tests effectively. The users are not software programmers or experts in the technology of the system. The software will provide an interface based on pictorial interaction and clear display of data and system status. Many interactive tools support these types of interaction, and the implementer might be expected to adopt existing software environments that would support such tools. An important component of the test-operator environment will be the control and organization of data collection. Data should be systematically collected and archived with integrated time references. Preliminary real-time display of data for all sensors in the system should be part of this environment.

Test-Protocol Development and Routine Maintenance

An onsite technical systems staff will be required to support PETMAN and to provide the capability to modify, design, and implement new procedures and tests. This programming environment should provide access to high-level definitions of mannequin behavior and motion and opportunities for users to change procedures, timing, and motions within a bounded set of options. The programming environment should have a strong visualization component. For the modification of behaviors and motions or the design of new tests, the visualization environment should provide animated views of candidate motions and evaluation measures related to forces, energy expenditure, temperature, and other factors. Simulation is an important element of this development process. In the review of background systems (see Appendix D), the simulation systems presented by American Android Corp. and Boston Dynamics were good examples of the types of simulation that might be embodied in the developers' programming environment.

Expert-System Update, Revision, and Debugging

The detailed internal programming code of the system should normally not be accessed in routine maintenance and operating procedures. However, there must be internal access by the developer or other designated support vendors to modify and update internal software code. Proper software engineering procedures and methods are critical and support any needed modifications at this level of detail. Independent testing and benchmarks of modular components of the hierarchic system should support systematic updating as needed.

CONCLUSIONS AND RECOMMENDATIONS

Conclusion 6-1: Taken independently, most of the PETMAN objectives are feasible. Collectively, they are likely to be in conflict and will require compromises.

Recommendation 6-1:

- Set priorities among the requirements to help to meet the objectives that have the highest payoff.
- Maintain close contact with the contractors, and jointly decide on design trade-offs.

Conclusion 6-2: The nontethered approach to mannequin design has merit. However, a nontethered design that meets even the objective level of the functional requirements of PETMAN would be difficult to achieve with state-of-the art technology.

Recommendation 6-2: Make the nontethered approach the ultimate target for mannequin design, but allow tethering for the initial stages. Phase the program to have a functional mannequin from the outset, and allow progressive enhancements that gradually minimize the tether until it is eliminated from the design.

Conclusion 6-3: A tethered mannequin design has the best potential to meet critical functional requirements. The tether may require special management to minimize its effects on mannequin movements.

Recommendation 6-3:

- Begin the design with a tethered approach that minimizes the size and bulk of the tether.
- Consider the tether attachment at the heel of a boot to minimize tether interference with the mannequin.
- From the onset of the design, observe tether management closely to minimize use of the tether.

Conclusion 6-4: A tethered approach burdens the design with the potential need for a tether-managing arrangement that follows the mannequin and distracts from its aesthetics and agility. Any design that reduces the bulk of the tether would help to minimize its disadvantages.

Recommendation 6-4:

- Package as much of the motion and control resources inside the mannequin as possible within the target 50th percentile male population envelope.
- Minimize the size of the tether by use of wireless communication and shared or complementary inputs and outputs; for example, cooling water from motors may used to heat the mannequin skin.
- Use dynamic braking of motors (which recharges an internal battery) to minimize the external energy supply and the size of the power cord in the tether.

Conclusion 6-5: The dynamic stability of marching in high step, jumping-jack exercises, and the low and high crawl with weapons is challenging and goes beyond the state of the art of humanoid technology. The PETMAN system could be powered and articulated to perform the exercises, but the controls may take more time to develop

Recommendation 6-5: Follow a phased approach to the development of PETMAN to attain more functionality as time goes by. Include all the power requirements in the initial phase, and add the controls as the technology and algorithms develop.

Conclusion 6-6: The system architecture takes the form of a hierarchic structure that integrates task definition and planning, and model- and behavior-based execution and representation.

Recommendation 6-6:

- Require a systematic modular approach to the design of both hardware and software systems.
- Require explicit linkage of system modules to maintenance, service, and test procedures for system monitoring and repair.

Conclusion 6-7: Software systems for PETMAN must support several classes of users to meet stated needs fully, including

- Operations testing and monitoring.
- Test development and routing maintenance.
- Expert programming.

Recommendation 6-7:

- A formal approach to the introduction of modern software engineering principles and practices should be an integral part of software development.
- Existing commercial software systems and tools seem to be available to support the functionality anticipated in this system and should be used to support maintenance, reliability, and upgrading of the resulting system.

7

A Complementary Approach to Meeting PETMAN System Goals

During the course of this study, the complexity of designing and constructing a PETMAN system became evident. Therefore, feasible and complementary avenues involving existing technologies were discussed that could meet many of the requirements specified for PETMAN in a short period and at much lower cost. In this chapter, we discuss one such feasible approach that uses a sensor-integrated body suit under the soldier individual protection ensembles (IPE) that would provide substantial improvements over current testing methods. This approach meets many of the PETMAN requirements, but does not provide for testing IPE in a live chemical agent environment.

The current Man-in-Simulant Test (MIST) uses soldiers rather than mannequins; this offers the benefit of real human movements. Recreating human movement, respiration, perspiration, and body proportions in a robotic system is a difficult and expensive proposition. One of the failings of MIST is its method of leak detection. Participants are outfitted with passive collection systems that provide a total exposure value with no spatial resolution to associate a leak with any particular movement and no periodic sampling for temporal analysis over the two-hour test duration. Another failing of the MIST is the inability to use an actual agent. However, if gases and vapors of equal molecular size and concentration diffuse and penetrate at equal rates, an appropriate simulant will diffuse through a breach in the suit and compromise its integrity in a manner similar to that of an actual agent.

THE SENSOR-INTEGRATED BODY SUIT

As discussed in Chapter 5, smart textiles or wearable electronic systems provide a possible platform for creating a real-time chemical sensing network.[1] There are also sufficiently mature technologies, to construct a whole-body integrated-sensor suit outfitted with real-time sensing of chemicals, ambient and body temperature, humidity, and such vital signs as heart rate and electrocardiographic readings in a stand-alone or self-contained mode.

The sensor-integrated body suit (SIBS) would meet the PETMAN requirements in Box 5-1 similar to the PETMAN soft skin (see Chapter 5), but would be worn by soldiers on the body and under the IPE. It could have a base fabric of typical textile materials used in military underclothing, such as Meraklon® (staple polypropylene fiber), or cotton, polyester, and blends and could be made to be form-fitting with spandex (see Chapter 5). Optical fibers can be integrated into the ensemble to take advantage of optical chemical-sensing technologies (see Chapter 3), which have demonstrated real-time detection of target chemicals that meets the optimal PETMAN concentration requirements. Although, it is likely that additional effort will be needed to develop the technology for this test application. Figure 7.1 is a diagrammatic representation of the form-fitting SIBS.

The IPE to be tested would be worn by the soldier over the form-fitting body suit, thereby ensuring that no changes are made to the IPE for the purpose of testing. With the SIBS, all the key requirements denoted in Box 5.1 can be met except the use of real chemical agents during the testing. Thus, a sensor-integrated approach can provide a real-world test environment and meet many of the PETMAN requirements.

Advantages of the SIBS

The proposed complementary approach will allow for a significant improvement in MIST without the expense and associated risk associated with a fully developed PETMAN at a small fraction of the costs in money and time. Using the SIBS will have the added advantage of real-time monitoring of temperature and humidity and of the vital signs of the test subject. Testing would incorporate time-stamped video recording to correlate body movements and initial IPE locations with all measurements accurately.

The SIBS would allow testing for all sizes of soldiers, not just the 50th percentile male measurements specified in Section 3.2.2 of the PETMAN requirements document. In addition, using actual soldiers will eliminate the

[1]Service, R. F. 2003. News Focus Technology: Electronic Textiles Charge Ahead. *Science*, 301(5635):909-911; Park, S., and S. Jayaraman. 2003. Smart Textiles: Wearable Electronic Systems. *MRS Bulletin* (August 2003):585-591.

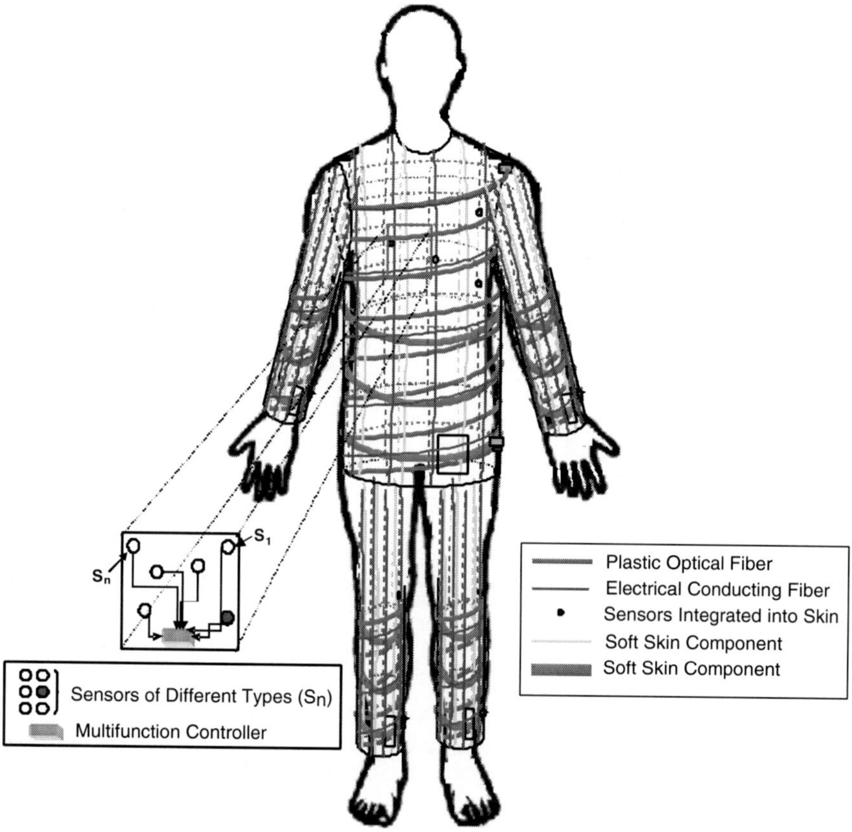

FIGURE 7.1 A form-fitting sensor-integrated body suit (SIBS) on a 50th percentile male.

cost and risk associated with robot development. Only a human moves like a human. The cost and technical difficulty of accurately mimicking human physiology are daunting. The proposed sensor-integrated approach would allow real-time monitoring and a huge variety of soldier movements and so provide added validation of suit integrity. This approach also eliminates the cost and potential hazards associated with the use of live agents.

Disadvantages of the SIBS

A potential drawback of this complementary approach is the inability to use live agents. If the test is designed to detect breaches in suit integrity, a live agent is not necessary. However, if the test is designed to assess the

TABLE 7.1 A Comparative Assessment of PETMAN and Complementary Approach

	PETMAN	SIBS on Human
Simulation of human movement	Poor and expensive	Excellent; minimal limitations
Performance in extreme environments	Minimal limitations	Human-like limitations
Use of actual agents	Yes	No
Problems with actual agents	Yes; decontamination required; potentially high cost if mannequin is compromised	Yes; cannot use; no requirement for specialized facility
Determination of breaches of IPE that result in chemical penetration	Yes	Yes
Simulation of human physiology	Poor	Excellent
Simulation of multiple soldier sizes	Single size, 50th percentile male	Available for all soldiers
Standalone (self-contained) use	No	Yes

behavior of an IPE in a live chemical environment, this requirement will not be met with the SIBS. Reproducibility and repeatability of defined motions are two key advantages of a robotic system. The SIBS would be worn by soldiers, and this might increase the acceptance of the test results in the soldier community, but the inherent variations introduced by human involvement could be another drawback of the proposed complementary solution.

A comparative analysis of PETMAN and the proposed complementary approach of the SIBS is provided in Table 7.1.

CONCLUSION

The SIBS approach would meet a large majority of the objective testing requirements at a small fraction of the cost of PETMAN. If there is an absolute requirement for a PETMAN robot, the sensor-integrated approach could provide many of the key capabilities in the near term. This complementary approach would offer substantial improvements over current testing methods while the critical paths for PETMAN development are being explored.

8

Overarching Conclusions and Recommendations

The feasibility of meeting the Protection Ensemble Test Mannequin (PETMAN) system threshold and objective requirements were considered individually and as an integrated whole. Specific conclusions and recommendations were presented in the preceding chapters of this report. Here, three overarching issues are presented for Product Director, Test Equipment, Strategy, and Support (PD TESS) to consider as it seeks to develop a PETMAN system: priority-setting among PETMAN requirements, contractor qualifications, and a complementary option to a PETMAN system. Each is discussed below.

PRIORITY SETTING AMONG PETMAN REQUIREMENTS

The Committee on Full-System Testing and Evaluation of Personal Protection Equipment Ensembles in Simulated Chemical-Warfare Environments was charged with assessing the feasibility of a PETMAN system on the basis of all delineated Department of Defense (DOD) system design requirements (see Appendix B). The PETMAN requirements cover all the desired and required features of mannequin function with little or no priority-setting. PD-TESS indicated that a concurrent exercise was being carried out to set priorities among the requirements but that the results would probably not be available in time to inform the PETMAN feasibility study. The feasibility study concludes the following:

Conclusion 8-1: Taken independently, most of the PETMAN threshold requirements can be met with current technologies or incremental developments of existing technologies (see detailed discussions in chapters 2 through 6).

Conclusion 8-2: Several options are available for chemical-agent sensing, robotic actuation, and overall system design.

Conclusion 8-3: Using currently available technologies, it may be possible to meet some of the threshold requirements in a nontethered system, but largely due to available battery technology such a system would be limited to an operating time of two hours. The other threshold and objective requirements may be difficult or impossible to meet with a nontethered mannequin.

Conclusion 8-4: Integrating all the objective requirements will be a major challenge for design and implementation of a full PETMAN system.

Conclusion 8-5: Meeting the threshold requirement of a tethered system, which would reduce the number of subsystems housed in the mannequin, is feasible.

In light of those conclusions, PD TESS should take the following actions:

Recommendation 8-1: To achieve greater success with the future proposal solicitation for a PETMAN system, PD TESS should set priorities among the PETMAN requirements according to the program objectives.

Recommendation 8-2: PD TESS should adopt a phased approach to the development of the PETMAN system, first addressing the high-risk areas as identified in the study (see Chapter 6) and then deciding on the achievable objectives according to the priorities it has set.

CONTRACTOR QUALIFICATIONS

The development of a PETMAN system is a large undertaking for any organization. The development of individual components—in particular the robotics capability—will require considerable resources and expertise. Simulating human physiology in addition to developing a robot may be beyond the means of a single group. On the basis of that assessment, the study concludes the following:

Conclusion 8-6: The design and development of a PETMAN system will require a multidisciplinary effort that encompasses expertise in computer software engineering, robot design, mannequin design, materials science and engineering, human physiologic simulation, sensor technologies, and systems integration.

PD TESS should take the following action in connection with assessing the qualifications of the company or companies chosen to develop a PETMAN system:

Recommendation 8-3: The primary contractor should have demonstrated capabilities in systems integration.

Recommendation 8-4: A workshop should be organized to inform the proposing groups fully of the objective and threshold requirements. The invitation list should include system integrators and developers and suppliers of component technologies for the mannequin, materials, and sensors.

A COMPLEMENTARY APPROACH

The current Man-in-Simulant Test (MIST) protocol evaluates individual protection ensembles (IPE) on soldiers rather than mannequins, and this offers the benefit of testing the effects of actual human movements and physiologic conditions. However, one of the major shortcomings of MIST is its method of under-ensemble data acquisition. It has been shown (see Chapters 2, 5, and 7) that recreating human-like movement, respiration, perspiration, and body proportions will be difficult and expensive. In contrast, simulant chemicals mimic almost all the physical features of actual chemical agents. On the basis of that assessment, the study concludes the following:

Conclusion 8-7: Some of the technologies reviewed in this report—such as real-time sensing of chemicals, temperature, and humidity—could be used in the MIST to provide real-time leak detection and characterization of the microenvironments in the protective garments.

Conclusion 8-8: Some technologies are sufficiently mature to support construction of a whole-body suit for a human—a sensor-integrated body suit (SIBS)—outfitted for real-time sensing of chemicals, body temperature, heart rate, cardiographic characteristics, and humidity without the need for a tether (see Chapter 7 for discussion of the SIBS). Such a suit would allow substantial improvement in the MIST without the expense and risk associated with a fully developed PETMAN system and at a small fraction of costs in money and time. The only limitation would be the inability to use actual chemical agents.

In light of those conclusions, PD TESS should take the following action:

Recommendation 8-5: A SIBS should be seriously considered as an update of the MIST and as complementary to the proposed PETMAN system. Unless there is an absolute requirement of all the capabilities associated with the proposed PETMAN system, this sensor-integrated approach would provide many of the key capabilities in the interim. Such an approach would provide substantial improvements over current testing while the critical paths and absolute requirements for PETMAN development are explored. It would allow testing to continue with human subjects and allow collection of data on a broader array of human characteristics than will be possible with the PETMAN system.

In light of the full list of requested functionalities of the PETMAN system, it will be difficult to design a nontethered, free-standing, 50th-percentile-male robotic test mannequin that can operate continuously beyond the available battery capacity of two hours. Tests longer than two hours will require reduced motion, a recharging or refueling method, a tether, or other approaches that would require research extending beyond the sponsor's desired development timeframe. The main design challenge for the PETMAN system is the integration of all system components (power, control, sensors, perspiration, respiration, actuation, and so on) within the size constraints. Setting priorities among the PETMAN systems requirements, improving guidance for proposing contractors, and considering complementary test approaches with simulants and real-time sensing will enhance the ability of PD TESS to develop a PETMAN system and ultimately improve the protection capability of IPE against chemical-warfare agents.

Appendix A

Statement of Task

The National Academies will assess the feasibility of full-system testing and evaluation of personal protective equipment ensembles in chemical and biological warfare agent environments under the conditions of simulated war fighter activities via an articulated, robotic mannequin. It shall

- Determine the feasibility of a Protection Ensemble Test Mannequin (PETMAN) system, based on all delineated Department of Defense (DOD) system design requirements for such a system.

- Focus on the significant design challenges associated with the DOD PETMAN system and whether and how they might be addressed.

- Discuss the cost-benefit and risk-benefit trade-offs associated with various design approaches to a PETMAN.

- Discuss whether and how some or all of the necessary protective ensemble test capability could be obtained if a PETMAN capability is infeasible. Discuss the cost-benefit and risk-benefit trade-offs of these alternatives.

Appendix B

Description of the PETMAN System Feasibility Study

1.0 SCOPE

1.1 GENERAL

The Joint Project Manager for Nuclear, Biological, and Chemical Contamination Avoidance (JPM NBCCA) was appointed by the Deputy Under Secretary of the Army for Operations and Research (DUSA (OR)), to identify gaps in the chemical and biological (CB) defense test and evaluation (T&E) infrastructure that may cause a shortfall in the ability to test emerging equipment. As part of accomplishing this responsibility, the Product Director, Test Equipment, Strategy and Support (PD TESS) has been established to manage the effort to identify T&E infrastructure gaps, align supporting efforts, and leverage all T&E assets.

PD TESS has identified the need for a T&E capability that will provide full system evaluation of protective ensembles in CB warfare agent environments. The Protection Ensemble Test Mannequin (PETMAN) system is being developed to test the protection capability of Individual Protection Ensembles (IPE) against CB warfare agents. The PETMAN system will be an articulated robotic mannequin that will perform exercises that simulate war fighter activities. The PETMAN system will be heated to produce body temperatures, able to perspire, and breathe.

This Performance Work Statement (PWS) defines the task requirements for a feasibility study for a PETMAN system design. The intent is to use the feasibility study as a tool to trade off requirements versus cost to support a future design effort.

1.3 GOVERNMENT FURNISHED EQUIPMENT / MATERIALS / FACILITIES

1.3.1 PD TESS will provide individual protective equipment technical manuals (TMs) to facilitate compatibility of the PETMAN system design to individual protective and ancillary equipment defined in **3.2.9-3.2.10.4**.

2.0 APPLICABLE DOCUMENTS

The following documents are referenced for the performance of this effort:

1. ***Test Operations Procedure (TOP) 10-2-022, Chemical Vapor and Aerosol System-Level Testing of Chemical/Biological Protective Suits.***
2. ***DOD-HDBK-743A, Military Handbook Anthropometry of U.S. Military Personnel, 13 February 1991.***
3. ***Fit Factor: M41 PATS technical manual/Protection Factor for equipment***
4. ***Weapons firing information: FM 3-22.9, Rifle Marksmanship M16A1, M16A2/3, M16A4, and M4 CARBINE, April 2003.***
5. ***DA PAM 385-61, Toxic Chemical Agent Safety Standards, 27 March 2002.***

3.0 REQUIREMENTS

3.2 PETMAN System Feasibility Study. The feasibility study will be based on the PETMAN system performance requirements defined in **3.3.1-3.3.14**. The focus of the feasibility study will be on the significant design challenges defined in **3.2.1-3.2.8** of the PETMAN system design in the feasibility study, however <u>all</u> requirements must be considered while conducting the study. The feasibility study will determine the requirements trade-offs and cost differentials associated with each design approach considered in the feasibility study.

PETMAN system performance requirements include threshold and objective requirements. Threshold and objective requirements are identified by (T) and (O), respectively. A threshold requirement denotes a minimum acceptable performance requirement while an objective requirement denotes the desired operational capability to meet system performance goals. If a requirement is not identified by (T) or (O), then the requirement is to be considered a threshold requirement.

The PETMAN system is defined as all components to include hardware and software required for operation, maintenance and calibration.

3.2.1 The study will determine the feasibility of designing a PETMAN system to be tethered (T) or free standing/self-contained (O). A tethered PETMAN system design must not compromise the integrity of the individual protection ensemble equipment being tested on the PETMAN system. If a tethered design is selected the design must also minimize the impact to the whole ensemble operation.

3.2.2 The study will determine the feasibility of designing a PETMAN system to be compatible with all individual protection and ancillary equipment as well as weapon systems defined in **3.3.9-3.3.10.4**. Areas to be addressed are donning/doffing and proper size/fit of the individual protection equipment. The PETMAN system design shall meet the appropriate 50th percentile male anthropometric measurements, as defined in DOD-HDBK-743A, Military Handbook Anthropometry of U.S. Military Personnel, to allow for the necessary fit/seal that each piece of protective equipment requires.

3.2.3 The study will determine the feasibility of designing a PETMAN system whose materials of construction will not be significantly degraded by exposure to both traditional chemical agents (T) and Toxic Industrial Chemicals (TICs) / Toxic Industrial Materials (TIMs) (O) and that can subsequently be decontaminated to negligible levels without adversely affecting the operation of the PETMAN system as defined in **3.3.11**.

3.2.4 The study will determine the feasibility of designing a PETMAN system with an integrated under ensemble sampling system that will allow for the collection of agent breakthrough data in real time (1-second increments).

3.2.5 The study will determine the feasibility of designing a PETMAN system that can simulate fixed skin temperature (by body region), perspiration rate (by body region), and respiration rate (T) and a realistic variability in skin temperature, perspiration rates, and respiration rates based on the amount of physical activity/exertion (O) defined in **3.3.4.1-3.3.4.3.**

3.2.6 The study will determine the feasibility of designing a PETMAN system capable of operating in fixed environmental chamber conditions

(T) and a range of environmental chamber conditions (O) as defined in **3.3.6.1-3.3.6.5**.

3.2.7 The study will determine the feasibility of designing a PETMAN system that can perform the Man-in-Simulant Test (MIST) exercises defined in **3.3.8** versus all motions. The study will determine the feasibility of the PETMAN system performing the human-like movements utilizing the degrees of freedom (DoF) and considerations defined in **3.3.7**.

3.2.8 The study will determine the feasibility of designing a PETMAN system with fully articulated hands and feet that simulate human motion, the minimum amount of hand and foot articulation required for the PETMAN system operation and a partial level of hand and foot articulation.

3.3 PETMAN System Performance Requirements for Feasibility Study.

3.3.1 The PETMAN system shall be robotic and either tethered (T) or free-standing/self-contained (O). A tethered PETMAN system design must not compromise the integrity of the individual protection ensemble equipment being tested on the PETMAN system. If a tethered design is selected the design must also minimize the impact to the whole ensemble operation.

- Free standing/self-contained is defined as having no external support. All power, fluids, heating and other components for operation will be contained internal to the PETMAN system.
- Tethered is defined as having external supports, power requirements and telemetry connections.

3.3.2 The PETMAN system shall be capable of operation for twelve (12) hours prior to requiring operational maintenance, three (3) months prior to preventive maintenance and six (6) months prior to calibration. (T) The PETMAN system shall be capable of operation for twenty-four (24) hours prior to requiring operational maintenance, six (6) months prior to preventive maintenance and twelve (12) months prior to calibration. (O) Operational maintenance is defined as the required maintenance procedures to prepare the PETMAN system for each test trial, for example, filling a perspiration reservoir, changing agent samplers or decontamination before the next trial. Preventive maintenance is defined as a maintenance event performed prior to a failure in order to prevent its occurrence.

3.3.3 The PETMAN system shall meet the anthropometric requirements of the 50th percentile male IAW DOD-HDBK-743A, Military Handbook Anthropometry of U.S. Military Personnel, 13 February 1991.

3.3.4 The PETMAN system shall simulate the following environmental/ physiological conditions under the individual protection ensemble.

3.3.4.1 The PETMAN system shall simulate fixed skin temperature by body region (T) and more realistic variability in body surface temperature based on body region and the level of physical activity/exertion (O).

3.3.4.2 The PETMAN system shall simulate a fixed perspiration rate of 0.4 L/hr (T) and more realistic variability in perspiration rates (range 0.11 to 1.8 L/hr) based on the level of physical activity/exertion (O).

3.3.4.3 The PETMAN system shall simulate a respiration rate of 50 L/min (fixed tidal volume of 1.5 L & breath frequency of 33 breaths/min) (T) and more realistic variability in respiration rates (range 10 to 115 L/min with variable tidal volumes and breath frequencies) based on the level of physical activity/exertion (O).

3.3.5 The PETMAN system shall be compatible with current under-ensemble chemical breakthrough sampling technologies, procedures, and equipment as defined in Test Operations Procedure (TOP) 10-2-022, Chemical Vapor and Aerosol System-Level Testing of Chemical/Biological Protective Suits (T) and designed to enable integration with real-time (1 second increments) sampling technologies, procedures, and equipment (O). At a minimum, sampling locations shall be the same as those defined in TOP 10-2-022.

3.3.6 The PETMAN system operation shall not be affected by the following chamber environmental conditions.

3.3.6.1 Temperature: 90°F ± 2°F (T); –25°F to 125°F ± 1°F, measured every 5 minutes (O)
3.3.6.2 Relative Humidity: 80% ± 3% (T); 0-100% ± 1%, measured every 5 minutes (O)
3.3.6.3 Wind speed: 0-10 mph ± 10% (T); 0-161 mph ± 2 mph (O)
3.3.6.4 Pressure: 0.25 iwg chamber vacuum maintained ± 2%
3.3.6.5 Liquid and vapor chemical agents including: All nerve and vesicant agents, as well as the chemical simulants, triethylphosphate and methyl salicylate.

3.3.7 The PETMAN system shall be articulated and robotic such that it looks and moves like a human, to include aesthetics, its proportions, and how the joints respond to sudden movements. All movements shall simulate realistic human control. The following, **3.3.7.1-3.3.7.16**, are the minimum degrees of freedom (DoF) and considerations to be made for the PETMAN system to mimic human control. Additional DoF or joints can be used such that the PETMAN system can mimic human control and don/doff individual protection equipment as defined in **3.3.9.1-3.3.9.10**.

3.3.7.1 The upper body of the PETMAN system shall consist of a head, a neck, a torso, two shoulders, two upper arms, two elbows, two lower arms and two hands.

3.3.7.2 The head shall be able to both pan and tilt as required.

3.3.7.3 The neck shall provide a minimum of two DoF between the torso and the head. This will allow the head to look up and down and from side to side, or pitch and yaw.

3.3.7.4 The torso shall be the base component of the upper body. The basic shape of the torso shall be similar to a male human chest. The torso shall have a minimum of three DoF, being pitch, roll and yaw. The PETMAN system shall be enabled to bend over, tilt from side to side and swivel with respect to the frontal plane.

3.3.7.5 The shoulders shall be attached to the torso and will allow for a minimum of two DoF.

3.3.7.6 The upper arm shall be attached at the shoulder, and shall be designed such that it is able to move with a minimum of two DoF without restriction.

3.3.7.7 The elbow shall have a minimum of one DoF and shall provide an attachment between the upper and lower arms.

3.3.7.8 The lower arm shall extend from the elbow joint and shall have a hand mechanism at its distal end.

3.3.7.9 The wrist shall be able to move with a minimum of two DoF for wrist extension/flexion and abduction.

3.3.7.10 The hand shall be articulated to study the glove/coat interface (T) or to have an opposable thumb and four fingers (O). Each finger shall have a minimum of four DoF (O). The distal interphalangeal joint (DIP) and the proximal interphalangeal joint (PIP) shall each have one DoF (O). The metacarpophalangeal (MCP) joint shall have a minimum of two DoF due to flexion and abduction (O). The opposable thumb shall have a minimum of three DoF (O). The thumb shall have one degree of freedom for the interphalangeal (IP) joint (O). The thumb MCP joint shall have a minimum of two DoF due to flexion and abduction (O). Additional (objective) thumb DoF shall be due to flexion and abduction (2 DoF) of the trapeziometacarpal (TM) joint (O).

3.3.7.11 The lower body shall consist of the waist, two hip joints, two upper legs, two knees, two lower legs, two ankles and two feet.

3.3.7.12 The hip joint shall allow the upper leg to move with a minimum of two DoF.

3.3.7.13 The knee shall move with a minimum of one DoF.

3.3.7.14 The ankle shall move with a minimum of two DoF. The ankle shall allow the foot to both pitch and roll.

3.3.7.15 The ball of the foot shall move with a minimum of one DoF providing toe roll motion while walking.

3.3.7.16 Actuation of joints shall be smooth and efficient to mimic human motion. PETMAN system movement shall be free of back-lash.

3.3.8 The PETMAN system shall be articulated and robotic such that it can simulate the exercises defined in **3.3.8.1-3.3.8.13.**

3.3.8.1 Standing

3.3.8.2 Walking at 4.8 km/hr (3 mph)

3.3.8.3 Marching (12 in. high step) at 4.8 km/hr (3 mph)

3.3.8.4 Jumping Jacks: Start with feet together and hands at sides. Simultaneously step left foot to the side causing feet to be shoulder width apart and bring hands together palm to palm over head. Simultaneously return left foot to starting position (feet together) and hands back to sides. Repeat the exercise with the right foot. (T) Start with feet together and hands at sides. Simultaneously bring hands together palm to palm over head while jumping and landing with feet shoulder width apart. Simultaneously jump bringing feet back together and hands back to sides. (O)

3.3.8.5 Sitting

3.3.8.6 Move from standing position to squatting position and return to standing position

3.3.8.7 Reaching arms in all directions

3.3.8.8 Move from standing to laying prone and return to standing

3.3.8.9 Kneeling on one knee

3.3.8.10 Kneeling on both knees

3.3.8.11 Low crawl: Lay prone. Keep body flat against the ground. With firing hand, grasp weapon sling at the upper sling--swivel. Let the front handguard rest on forearm (keeping the muzzle off the ground), and let the weapon butt drag on the ground. To move, push arms forward and pull firing side leg forward. Then pull with arms and push with leg. Continue this throughout the move.

3.3.8.12 High crawl: Keep body off the ground by resting on forearms and lower legs. Cradle weapon in arms and keep its muzzle off the ground. Keep knees well behind buttocks so body will stay low. To move, alternately advance right elbow and left knee, then left elbow and right knee.

3.3.8.13 Aiming weapon in various positions: standing, kneeling on one knee, laying prone (grip rifle, sight rifle, trigger pull associated with small arms)

3.3.9 The PETMAN system shall be compatible with the individual protection and ancillary equipment listed in **3.3.9.1-3.3.9.11**. The PETMAN system shall be designed such that the individual protection equipment can be properly donned IAW the respective technical manuals.

3.3.9.1 Suits

3.3.9.1.1 Joint Service Lightweight Suit Technology (JSLIST) Type II
3.3.9.1.2 JSLIST Type VII
3.3.9.1.3 All Purpose-Personal Protective Ensemble (AP-PPE)
3.3.9.1.4 Chemical Protective Undergarment (CPU)

3.3.9.2 Boots
3.3.9.2.1 Green Vinyl Overboot (GVO) / Black Vinyl Overboot (BVO)
3.3.9.2.2 Multipurpose Overboot (MULO)

3.3.9.3 Gloves
3.3.9.3.1 7, 14, and 25 mil butyl gloves
3.3.9.3.2 JSLIST Block 1 Gove Upgrade (JB1GU) and JB1GU-Flame Retardant (FR)

3.3.9.4 Masks
3.3.9.4.1 M40/42 Series Masks
3.3.9.4.2 M45 Mask
3.3.9.4.3 M48 Chemical-Biological Apache Aviator Mask
3.3.9.4.4 Aircrew Eye Respirator Protection (AERP)
3.3.9.4.5 AP-22P Respirator Assembly

3.3.9.5 Helmets
3.3.9.5.1 Advanced Combat Helmet (ACH)
3.3.9.5.2 Personal Armor System Ground Troops (PASGT) Helmet
3.3.9.5.3 Modular Integrated Communications Helmet (MICH)
3.3.9.5.4 Lightweight Helmet

3.3.9.6 Combat Boots

3.3.9.6.1 Lightweight Desert Combat Boot
3.3.9.6.2 Jungle Boot
3.3.9.6.3 Infantry Combat Boot
3.3.9.6.4 Temperate and Hot Weather Combat Boot

3.3.9.7 Ballistic Protection Vests
3.3.9.7.1 Spall Vest
3.3.9.7.2 Interceptor Vest

3.3.9.8 Pistol Holsters

3.3.9.9 All Services Battle Dress and Combat Uniform

3.3.9.10 All Services Physical Training Gear (T-shirt, running shorts, socks)

3.3.9.11 Skin Exposure Reduction Paste against Chemical Warfare Agents (SERPACWA) (O)

3.3.10 The PETMAN system shall be compatible with the following weapons systems. The PETMAN system must be able to hold/grip and aim the weapon IAW the field manual (FM) FM 3-22.9, Rifle Marksmanship Field Manual.

3.3.10.1 M4 Modular Weapon
3.3.10.2 M24 Sniper Rifle
3.3.10.3 M16A2 Rifle 5.56 MM
3.3.10.4 XM8 Lightweight Assault Rifle

3.3.11 The PETMAN system shall be capable of being decontaminated with no adverse effects on the operation of the system and such that there is no effect on the next iteration of test (T) or leaving negligible agent residual, as defined by 3X decontamination level in DA PAM 385-61, Toxic Chemical Agent Safety Standards, (O) on the PETMAN system.

3.3.12 The PETMAN system shall utilize as many common commercially available parts and/or components as possible.

3.3.13 The PETMAN system shall record the following system parameters over time: skin temperature, respiration rate, perspiration rate, and total mass (in nanograms) of chemical vapor that penetrates/permeates through the protective ensemble. The PETMAN system shall record the start and stop time of each motion in 1 second increments.

3.3.14 The PETMAN system shall be programmable to perform a series of exercises or motion. The PETMAN system shall track body position during all motions within 1-second increments.

Appendix C

Committee Biographic Information

Masayoshi Tomizuka (*Chair*) is the Cheryl and John Neerhout, Jr., Distinguished Professor of Mechanical Engineering at the University of California, Berkeley and is a former program director for the Dynamic Systems and Control Program in the Civil and Mechanical Systems Division of the National Science Foundation. Dr. Tomizuka's research covers control theory and its applications to various mechanical systems, adaptive control, computer-aided manufacturing, control systems and theory, digital control, dynamic systems, manufacturing, and mechanical vibrations. Dr. Tomizuka received his B.S. and M.S. from Keio University in Japan and his Ph.D. from the Massachusetts Institute of Technology.

Hadi Abu-Akeel (NAE) is president of AMTENG Corp., an independent consulting firm in industrial manufacturing robots. Dr. Abu-Akeel recently retired from FANUC Robotics NA, Inc., an industrial robotics firm, where he was senior vice president and chief engineer. His main expertise includes optimization of robot design, including tradeoffs of performance, cost, manufacturability, application requirements, and user friendliness; use of robotic devices to overcome manufacturing productivity challenges and provide cost-effective manufacturing-process alternatives; development and application of microsensors for intelligent robots, robotic assist devices, autonomous robots, and remote presence; and risk assessment, safety, and safeguarding of robot applications. In 1997, he was elected to the National Academy of Engineering for contributions to design, control, and implementation of industrial robots. Dr. Abu-Akeel received his Ph.D. in mechanical engineering from the University of California, Berkeley.

Christopher G. Atkeson is a professor in the Robotics Institute and Human-Computer Interaction Institute at Carnegie Mellon University (CMU). He received his M.S. in applied mathematics (computer science) from Harvard University and his Ph.D. in brain and cognitive science from the Massachusetts Institute of Technology (MIT). He joined the MIT faculty in 1986, moved to the Georgia Institute of Technology College of Computing in 1994, and moved to CMU in 2000.

Lisa M. Brosseau is an associate professor in the School of Public Health at the University of Minnesota. She is an industrial hygienist whose primary interests are in controlling infectious aerosols and assisting small businesses with workplace health and safety. Dr. Brosseau received her master's degree and doctorate in industrial hygiene from the Harvard School of Public Health. Her current research focuses on testing the effectiveness of health and safety interventions in small metalworking shops and developing and testing the effectiveness of health and safety newsletters in small manufacturing businesses. Dr. Brosseau is deputy director of the Midwest Center for Occupational Safety and Health, which is supported by a National Institute for Occupational Safety and Health training grant. She teaches courses in applications of industrial hygiene and management of hazardous materials and waste.

Zane Frund is the manager of chemical research and materials engineering at Mine Safety Appliances (MSA, Inc.), where he is responsible for the development and evaluation of designs and compounds used in a wide array of occupational health and safety equipment applications (for example, air-purifying respirators, body armor, firefighter self-contained breathing apparatus, thermal imaging cameras, and solid oxygen-containing self-rescuers). Dr. Frund developed a curriculum and gave college and university courses associated with occupational health and respiratory protection, materials science and engineering, forensic science, and forensic chemistry. He has also served as a peer reviewer of technical reports and manuscripts for the National Institute for Occupational Safety and Health, the *Journal of Occupational and Environmental Hygiene*, and the *Journal of the International Society for Respiratory Protection*. He received his Ph.D. in materials engineering with a minor in occupational and environmental health from the University of Pittsburgh.

Darrell L. Jan is the project manager for the National Aeronautics and Space Administration (NASA) Advanced Environmental Monitoring and Control (AEMC) program element in the Exploration Systems Technology Office at the Jet Propulsion Laboratory, a federally funded esearch and development center staffed and managed by the California Institute of Technology. In this role, Dr. Jan is responsible for developing chemical-sensing systems for future human space flight. As manager of AEMC, he has overseen research projects that have included miniature mass spectrometry, an electronic nose array, development of mid-infrared tunable diode lasers for trace-gas measurement, polymer chair reaction and and adenosine tri-phosphate analysis of space-station water, detection of trace gases via fluorescence resonance energy transfer using dendritic molecules, bioluminescent bioreporter chips, and microfluidic ion chromatography. Dr. Jan has received a NASA Quality and Productivity Award for propulsion system filter analysis. He is a member of the American Institute of Aeronautics and Astronautics, the American Association for the Advancement of Science, the American Institute of Chemical Engineers, and the American Society of Mechanical Engineers. Dr. Jan earned his B.S. in bioengineering at the University of California, Berkeley and his S.M. and Ph.D. at the Massachusetts Institute of Technology in the Department of Mechanical Engineering.

Sundaresan Jayaraman is a professor of polymer, textile, and fiber engineering with a joint appointment in the College of Management at the Georgia Institute of Technology in Atlanta, Georgia. He and his research students have made substantial contributions in enterprise architecture and modeling methods for information systems, engineering design of intelligent textile structures and processes, design and development of knowledge-based systems for textiles and apparel, and multimedia educational systems. His group's research has led to the realization of the world's first Wearable Motherboard™, also known as the Smart Shirt. He received his BTech and MTech in textile engineering from the University of Madras, India, and his Ph.D. in textile engineering from North Carolina State University.

Leo Kobayashi is the director of adult simulations at the Rhode Island Hospital Medical Simulation Center. He graduated from Brown Medical School and completed his emergency-medicine residency at Brigham and Women's Hospital/Massachusetts General Hospital in 2002. Dr. Kobayashi is an active educator in Brown Medical School and its postgraduate training program in emergency medicine. His research focuses on advanced medical simulation, its validation as an educational method, and its application to disaster-medicine education and training. He has substantial experience in medical simulation in teaching duties at Harvard Medical School and at the Rhode Island Medical Simulation Center. He has helped in developing the concept of multiple patient simulations for emergency care and disaster response.

Dava J. Newman is a professor of aeronautics and astronautics and engineering systems and director of the Technology and Policy Program at the Massachusetts Institute of Technology (MIT). Her multidisciplinary research effort combines aerospace bioengineering, human-in-the-loop dynamics and control modeling, biomechanics, human interface technology, life sciences, and systems analysis and design. The research studies are carried out through flight experiments, ground-based simulations, and mathematical and computer modeling. Other research efforts include advanced space-suit design and navigation aids for extravehicular-activity astronauts. Dr. Newman earned her B.S. in aerospace engineering from the University of Notre Dame in 1986, dual M.S. degrees from MIT in 1989 in aeronautics and astronautics and technology and policy, and a Ph.D. from MIT in aerospace biomedical engineering.

Arthur C. Sanderson is professor of electrical, computer, and systems engineering, and vice president of research at Rensselaer Polytechnic University. After completing a postdoctorate at Delft University of Technology (TU Delft) in The Netherlands, Dr. Sanderson held faculty positions at CMU from 1973 to 1987, where he was codirector of the Robotics Institute. He also held visiting positions at Delft University of Technology; Universidad Iberoamericana, Mexico City; Philips Laboratories, Briarcliff Manor, NY; and the University of Tsukuba, Japan. In 1987, he joined Rensselaer Polytechnic Institute as a professor and served as head of the Electrical, Computer and Systems Engineering Department. During 1998-1999, he served as director of the Division of Electrical and Communications Systems at the National Science Foundation in Washington, D.C. From 2000 through July 2004, Dr. Sanderson served as vice president for research of Rensselaer Polytechnic Institute and was responsible for development and coordination of research programs on the campus. He received his B.S. from Brown University and his M.S. and Ph.D. from Carnegie Mellon University (CMU).

R. Paul Schaudies is chief executive officer of GenArraytion. Inc. GenArraytion is a veteran-owned biotechnology company that uses cutting-edge molecular-biology and bioinformatic techniques to identify and characterize pathogenic organisms. Previously, he was the assistant vice president and division manager at Science Applications International Corporation. His division focused on three major business areas: contract biomedical research, technology assessments, and scientific studies. He was key in establishing the levels for reentry into the Hart Senate Building (after it was contaminated with anthrax in 2001) and is a nationally recognized expert in biologic- and chemical-warfare defense. Dr. Schaudies served on numerous national advisory panels for the Defense Intelligence Agency (DIA), the Defense Advanced Research Projects Agency, and the Department of Energy. He has 14 years of bench research experience in managing laboratories at Walter Reed Army Institute of Research and as a visiting scientist at the National Cancer Institute. He served for 13 years on active duty with the Army Medical Service Corps and retired from the U.S. Army Reserves as a lieutenant colonel. He spent 4 years with DIA as collections manager for biologic and chemical defense technologies; he initiated numerous intra-agency collaborations that resulted in accelerated product development in biologic-warfare agent detection and identification. Dr. Schaudies has served on many National Research Council committees, including the Committee to Review the National Nanotechnology Initiative.

Appendix D

Open Session Presentation Summaries

The committee heard from several people in open data-gathering sessions over the course of its four meetings. Their presentations are summarized briefly below. Specific technologies mentioned are discussed in more detail in the relevant chapters of the report as indicated.

INDIVIDUAL PROTECTION ENSEMBLES TESTING

Charlie Walker, of the U.S. Army Dugway Proving Ground, provided an overview of the Man-in-Simulant Test (MIST) during meeting 2, including video of soldiers performing various tasks while wearing individual protection ensembles (IPEs). The MIST is discussed in more detail in Chapter 1.

HUMAN SIMULATION

Information provided by the speakers described in this section (who all presented during meeting 3) is also included in the discussion on current capabilities in Chapter 2.

Grant Bue, of Johnson Space Center, discussed National Aeronautics and Space Administration spacesuit technology testing (by telephone). Bue focused on approaches to under-suit cooling for astronauts, which was developed for space exploration. Descriptions of several feasible technologies applicable to PETMAN were discussed by Bue, including advanced heat pumps, a mini-vapor-compressor, "super ice," and a vortex tube.

Rick Burke, of Measurement Technology Northwest (MTNW), spoke about thermal mannequins for human-physiology simulation. Burke discussed theory and applications of MTNW equipment and capabilities including the Advanced Automotive Manikin (ADAM). He highlighted a few major PETMAN challenges and recommendations: Internal space in the 50th percentile body form is likely to be insufficient for all his combined systems, thermal management of the system will be necessary, simplification of joints compromises the realism of body contours, internal battery power and charging systems are large and cumbersome, the internal breathing and humidification system is limited by the size of the mannequin head cavity, and the internal fluid reservoir for sweating could be replaced with an external supply without affecting performance.

Carlos Moreno, of Medical Education Technologies, Inc. (METI), described METI's medical educational mannequins and human-physiology simulation capabilities. METI provides integrated physiologic models with the hardware to represent patient responses accurately. The main use of the models is in evaluation of student performance. One model in particular—iStan—is completely self-contained, has skin made of thermoplastic elastomer, has a realistic weight, and uses pressurized water to create sweating. Other features include an 8-hour operational mode and an area in the belly that is empty to allow for additional simulation capabilities.

SENSING CAPABILITIES

Information provided by the speakers described in this section is included in more detail as part of the discussion on current capabilities in Chapter 3. Each of the technologies has the potential to be the chemical-sensing component of a PETMAN system. However, these are just examples of technologies that would be available within 2 years of development for PETMAN. There are probably more technologies with similar potential that are not included here.

H. James Harmon, of Oklahoma State University, spoke (by telephone) at meeting 2 about the technology he has developed for real-time reagentless solid-state optical detection.

Committee member Zane Frund spoke with a representative of **Seacoast Science, Inc.,** between meetings 2 and 3. Seacoast Science produces chemicapacitor-based chemical sensors.

David Walt, of Tufts University, spoke at meeting 3 about his fluorescence-based optical microsphere arrays and fiber-optics chemical-agent sensing capabilities.

ROBOTICS CAPABILITIES

More detailed information provided by the speakers described in this section, who all spoke during meeting 3, is included in the discussion on state-of-the-art capabilities in Chapter 4. A number of other robot systems are also discussed in Chapter 4.

Robert Ambrose, of the National Aeronautics and Space Administration (NASA) Johnson Space Center, provided an overview of Robonaut (by telephone). Ambrose presented the history of NASA's Robonaut, details on the anatomy of the first-generation Robonaut R1, videos demonstrating capabilities, and plans for development of the next-generation Robonaut R2. Robonaut has human-like dexterity but moves much slower. Currently, Robonaut features only a human-like upper body. The development of Robonaut over the last decade cost about $25 million.

David Handelman, of American Android Corporation, spoke about all-terrain biped technology, focusing on modeling capabilities.

Jun Ho Oh, of Hubo Lab, the Humanoid Robot Research Center in the Department of Mechanical Engineering of the Korea Advanced Institute of Science and Technology, provided an overview (which included many detailed specifications) of human-like machines and types of actuators and presented a case review of Hubo and comments and suggestions about PETMAN. Oh summarized by saying that the performance requirements for PETMAN are too severe and comprehensive. Oh suggested that movement tests be conducted separately from static tests, pneumatic actuators be used for the dynamic test, head and finger and foot tests be performed separately, and power and air sources be external to the system.

Robert Playter, of Boston Dynamics, discussed its robotics capabilities in relation to PETMAN requirements. Playter provided a company overview. He also discussed control systems for human movement and robot behaviors; power requirements for legged locomotion based on measurements from "BigDog," including energy and heat estimates; and systems-design issues. He highlighted four "tall poles" or major challenges of PETMAN—power and heat, control systems, energy, and mechanical design. Playter also discussed the various control options (electric, hydraulic, and pneumatic) and said that he does not believe that PETMAN can be built without a tether.